ENCYCLOPEDIA OF THE UNEXPLAINED

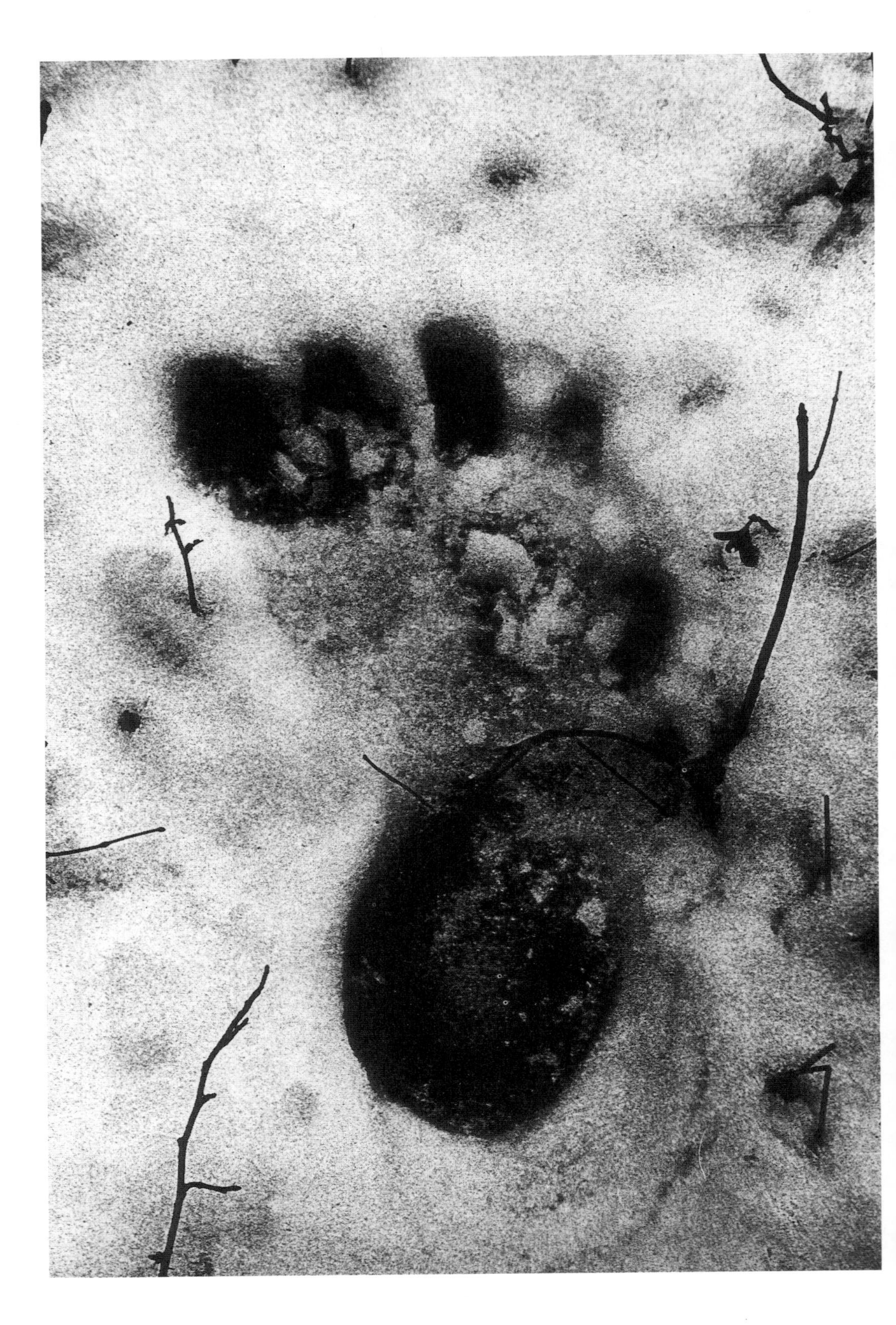

ENCYCLOPEDIA OF THE UNEXPLAINED

Jenny Randles and
Peter Hough

BROCKHAMPTON PRESS
LONDON

First published in Great Britain in 1995 by
Michael O'Mara Books Limited

This edition published in 1998 by
Brockhampton Press
a member of Hodder Headline PLC Group
ISBN 1-86019-409-5

A CIP catalogue record for this book is available from the British Library

Designed by Mick Keates

Typeset by Florencetype Ltd, Stoodleigh, Devon
Printed in Slovenia by Tiskarna Mladinska knjiga

People who wish to report any unusual phenomena can write to the authors – in complete confidence if preferred – care of 11 Pike Court, Fleetwood, Lancashire SY7 8QF, UK.

Frontispiece: Footprint of the so-called 'Bossburg cripple', found in the snow at Bossburg, Washington in 1969. It measured over 16½ inches long and appeared deformed – hence the name. Man-beasts have frequently been sighted in this north-west area of the United States (see p. 182)

CONTENTS

PART ONE: THE SUPERNATURAL EARTH

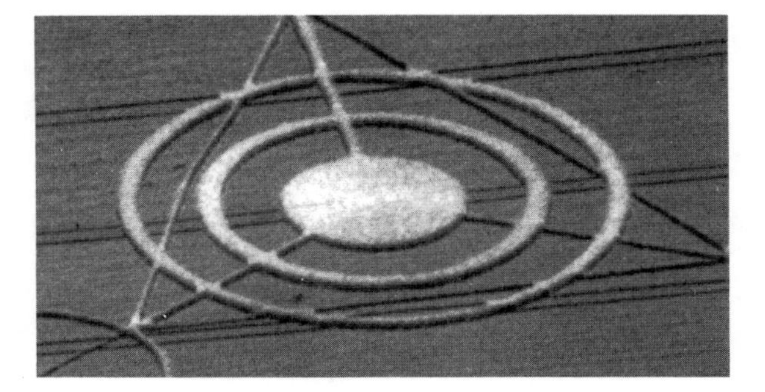

PART TWO: SPACE INVADERS

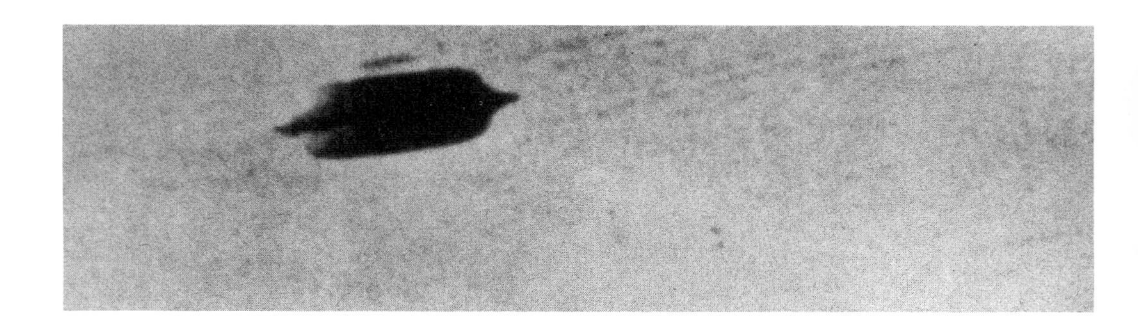

PART THREE: OUT OF TIME

PART FOUR: DEATH BY SUPERNATURAL CAUSES?

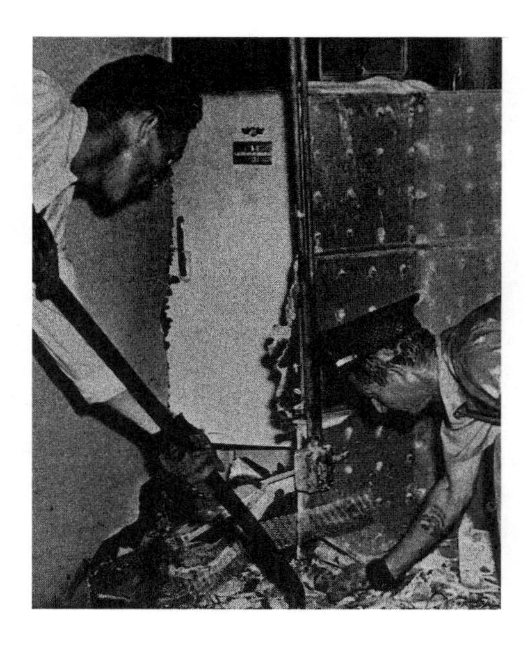

PART FIVE: MIND MATTERS

PART SIX: THE SPIRITUAL DIMENSION

Religious Miracles have been alleged for thousands of years. Most religions set great store by them. Yet even in our technological age the modern-day equivalent of turning water into wine is claimed with surprising frequency. Statues move, blood and tears pour from religious icons and mysterious pictures appear from out of nowhere. Do they reflect a deeper level to both mind and body – a spiritual dimension to human life?

126

Apparitions is a subject that everyone talks about. We enjoy being scared by a ghost story. Yet there are many curious facts about these phantoms that are seldom discussed. Here we explore some of them. We look at alternative ways of understanding the nature of a ghost beyond the more obvious possibilities of the dead coming back to earth. These introduce startling scientific possibilities for the future, such as the amazing prospect of watching the Last Supper 'live' on your TV set.

132

The Near-Death Experience takes us as close as we can get towards the ultimate question – does any part of us survive death? Thanks to the wonders of modern medicine, people are regularly being snatched back from what, until very recently, would have been an irreversible demise. Yet many come back with a fantastic story to tell. These tales are weirdly consistent, giving first-hand accounts of what it felt like to die and what was waiting for them on the other side. Science is facing up to the challenge with a vengeance, knowing that in the near-death experience may lie the ultimate proof of a world beyond this or of the utter gullibility of the human mind.

143

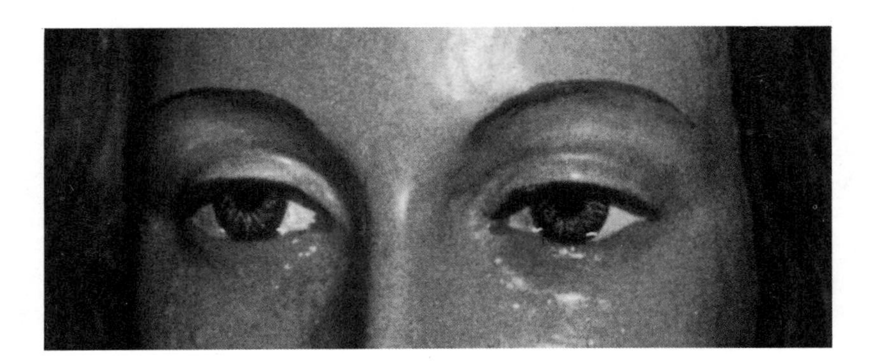

PART SEVEN: STRANGE BEINGS

Fairy Folk have long been considered denizens of another dimension that exists parallel to our own. Of course, there are old wives' tales from the days before the triumph and enlightenment of science. Nobody today should believe in fairies . . . yet, they do. The evidence can be examined with surprising objectivity. Will it reveal that we do, after all, share this world with supernatural beings?

156

The Men in Black are latter-day equivalents of goblins and demons. Sinister beings who plague witnesses after strange phenomena, offering threats, asking bizarre questions, then vanishing into nowhere. They are rarely discussed, yet are far more widely seen than generally realized. The Men in Black are an important part of the deeper mysteries of the supernatural world.

166

Phantom Visitors are an example of a paranormal phenomenon in embryo. Houses have been visited by strangers claiming to be social workers and making extraordinary demands. Yet, when investigations are carried out the visitors do not exist. They have melted into nowhere. Cases are very recent and are yet to be recognized as a supernatural event in some quarters. But they offer the unique opportunity to see how a brand new mystery begins, starts to develop and takes on a life of its own.

175

PART EIGHT: THE ALIEN ZOO

Man Monsters are like living examples of the missing link. Science has heard stories of them for many years and has sought hard for proof of their reality. There is some proof but why has no ape-man been captured? Are these giant creatures real flesh-and-blood animals that are simply very rare? Or is there a more supernatural explanation?

Sea Serpents and Monsters of the Oceans reports are as old as the sea itself. All over the world mariners claim to have seen frightening beasts that seem to be left behind from the primeval past. But are these stories simply legends, spun to impress on people that the sea must be respected? Or is there substance behind the shadows? Over two-thirds of this planet's surface is covered by water and we know less about its darkest depths than we do about the moon.

The Loch Ness Monster. Loch Ness hosts the world's favourite monster which has been given the name 'Nessie' to make it friendly as well as saleable as a tourist commodity. But what truth lies behind the countless reports of a strange giant creature lurking in this inky Scottish lake? We assess the strengths and the weaknesses of the extensive evidence.

Living Dinosaurs. These creatures have intrigued mankind for centuries. Officially they became extinct suddenly some 60 million years ago. The world is full of surprises for we are still finding creatures that we did not know existed and others we thought had long since disappeared. Dinosaurs may also live on. There are reports of survivors on land and in the air and several expeditions have already been mounted to find them within their own lost world . . . the results have been amazing.

PREFACE

This is a book full of mysteries. But unlike most enyclopedias or general surveys about the world of the unexplained we have chosen not to be superficial or to gloss over problems. Rather than write a few words about a hundred subjects, many of which are of dubious reality, we have instead focused on just a handful of the true mysteries of modern science – covering the Earth, space, time, mind, body, spirit and the animal kingdom and discussing each in the sort of detail that they deserve.

We have tried to do this objectively. This is not a book that sets out to prove reality above unreality. We aim merely to give you the facts, the arguments, the latest evidence and detailed references so that you may check them all for yourself. This is a source book that you can use in your own way. It is a book dedicated to telling things as they really are without the need to make you believe or disbelieve.

We think that you deserve to be treated with respect and to face up to these phenomena as they stand, warts and all. Then you can make up your own mind about their nature.

Good hunting!

Opposite: Bigfoot investigator, René Dahinden, measures 15-inch Bigfoot tracks found on Blue Creek Mountain in California, USA in 1967 (see p.182)

THE SUPERNATURAL EARTH

Stonehenge, the famous earth-mystery site in Wiltshire, England, is probably 10,000 years old. Its significance has taxed the minds of many researchers but it is the focal point of powerful forces and great reverence

LINES OF POWER

When is a line not a line?

There are a number of popular myths associated with the supernatural. One of these concerns the name of the infamous ship from which an entire crew mysteriously vanished. It is commonly referred to as the *Marie Celeste* but it was really called the *Mary Celeste.* Such minor errors or misconceptions can generate all manner of legendary assumptions which grow as time goes by.

Another misconception is the mysterious phenomenon of 'ley lines'. In truth, such terminology is redundant because the word 'ley' means line and so the correct application is to simply call them 'leys'. The word was adopted by Alfred Watkins early in the twentieth century when he 'rediscovered' alignments between ancient monuments, very old sites and places of religious significance scattered around England. His book *The Old Straight Track* became the genesis for today's field of 'earth mysteries', where all such matters are explored.

Discovering leys involves a combination of detective work with ancient maps, walking around the sites to check out the marker points that may or may not be identified, and an element of trial and error. Statistics indicate that you can draw random straight lines through any small combination of points sprinkled over a map. As a result great caution is needed to be sure that there are enough marker points of clear significance on the discovered ley to rule out chance effects.

This pursuit of 'ley hunting' has resulted in an alleged network across Britain and also other parts of northern Europe, such as Scandinavia, Ireland, France and Spain. Some argue that ancient people, certainly neolithic and perhaps the Celts, originally plotted the lines for obscure reasons and built burial mounds and much later churches as wayposts to mark them out. At some intersection points of major ley systems truly impressive monuments, such as the Avebury stone circle, seem to have been created.

Many place names in England have the root 'ley' within them. While this sometimes seems to derive from other meanings (e.g. 'lea' – or field) it is considered probable by some researchers that locations that can trace the term 'ley' far back may at times owe this to their supposed placement along a line of power. Historical research into odd events that may have occurred in the area for centuries is sometimes productive.

Aside from the artificial monuments, marker posts, or other factors, many of which have vanished beneath the onslaught of modern civilization, the leys themselves are usually impossible for the untrained eye to see, even on modern maps. They rarely have any clear visible indication such as an old bridle path or coaching road might have.

Precisely what these lines are, why they were considered important and what the significance or power of the impressive constructions still found at some nodal points is are all topics of continuing fascination to earth mystery students. For some people the answer lies in nothing strange or mysterious, simply the ability to try to tune into the thought processes and traditions of the original founders of north European civilization. But to others the leys may be the key to unlock a real or physical energy within the Earth itself which can be used to the benefit of mankind if its strength can be harnessed.

The Stonehenge Key

Tourists from all over the world visit the circle of standing stones on Salisbury Plain known as Stonehenge. Druids hold ceremonies there. New Age mystics commune with the spirit of the Earth felt within its heart and researchers of the supernatural have tried everything from astronomical observations to readings of electrical and magnetic energy radiated by the rock. As to the original meaning and purpose of the place theories continue to vie with one another for supremacy.

The rocks from which this impressive construction were built are not local. In fact they were brought over 100 miles from South Wales. Inevitably there must have been a reason for undertaking this wonder of engineering. Why not use local stones or, indeed, simply build Stonehenge *in situ* where the rocks were found? It suggests that there is some significance in the precise type of rock, with its quartz content, and the location of the circle on Salisbury Plain.

The site of Stonehenge has been constantly re-used. Carbon dating of holes in the ground here indicate that an earthen circle was probably first laid out as long ago as 8000 BC. For millenia it was added to and updated. The stone circle itself seems to be at least 5000 years later than the first deification of the area. As such it may have had many different purposes. A common belief is that it is an astronomical observatory, perhaps an aid to judging the seasons for agricultural purposes. There were some complex relationships between the position of the sun and stars some 5000 years ago and the position of some of the stones, suggesting they might have been markers. The Druid ceremonies which followed later seem to have used Stonehenge for that reason but there could well have been other religious uses.

Much more recently some new theories about the meaning of Stonehenge, in particular, and other circle sites have been proposed. One idea is based on a computer simulation carried out in California, USA, where it was found that 'mock suns' or mirage effects created by the sunlight shining on ice crystals in the atmosphere have a set mathematical relationship. This matches surprisingly well with the position of some of the key stones in these circle sites. Were the occasionally visible lights in the sky produced by this effect thought to be godly wonders and their mathematical positions mapped out in reverence?

Even more radical is the theory proposed in 1992 by a meteorologist, who argued that there was a reason why the stone circles resembled modern-day crop circles (*see* page 28) and also concentrated in the same area as these. He argued that the force that produced the crop circles (he believed an aerial vortex) was thought by early man to 'seed' the earth and 'penetrate' fertile crops, producing the circles. The ancients chose to build lasting testaments to this heavenly message in their great stone circle temples. The concept has had a mixed reception from earth mystery students and archeologists.

The ley of the land

Nobody is quite sure why Alfred Watkins called these alignments 'leys'. The most common theory is that he did notice how often they passed through places with 'ley' names. These tended to have deeply mystical significance, for example, Alderley Edge in Cheshire, where many tales of wizardry are told and local legends claim that King Arthur's spirit rests there waiting to save England in its hour of need.

However, this view is complicated by the fact that there appear to be ley systems in other countries. A good deal of research in Germany has uncovered straight paths leading to very old burial sites. As a result a new idea was pioneered by Paul Devereux, editor of *The Ley Hunter* magazine, regarding 'Spirit paths', identified as stretches of ancient track way. He believes that rather than interpret them as lines of energy created by the 'biosphere' of a living planet, they are trails along which sensitive people felt 'drawn' towards a spiritual centre, today often demarked by a church.

Devereux thinks that there is an interrelationship between much of the earth mysteries' field and the collective consciousness of human beings. In the past we were far more in tune and all tribal societies, even today, have shaman or medicine men who are greatly respected by their peers. They seem capable of communing with a different level of reality and there are some grounds for likening them to what our technological society calls psychics.

Aborigines, the native Australians, for example, put great store by the image of a 'dream

A ley is a straight line linking sites of earth-mystery relevance, such as stone circles and very old churches. The hunt for leys is a popular activity in Britain but their purpose remains controversial

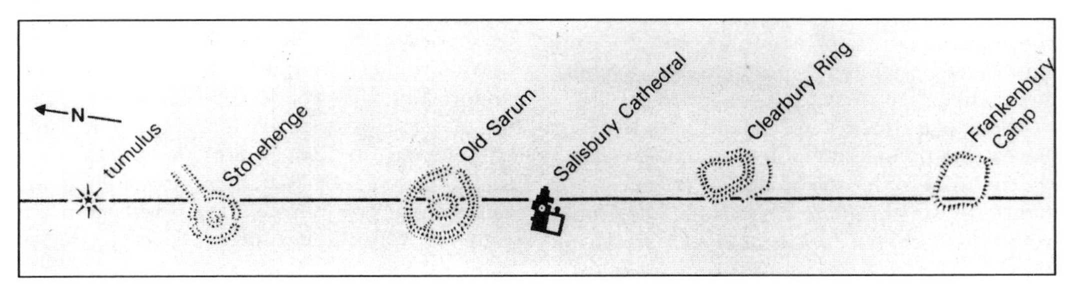

These mysterious lines are etched into the dust and rock of the Nazca plains of Peru, some forming shapes, including those of animals. It is known that they were created across hundreds of years but their meaning is unknown

time' within which their ancestors repose and where normal time and space no longer applies. They have a deep tradition of a serpent that sleeps within this dream time and is an intimate part of the Earth's spirit. One has to avoid awakening it or disasters can result, but those in touch with the serpent spirit can follow it towards better things.

There are known cases of Aboriginal shaman leading whole tribes of people on long treks. It has been said that they have led them from areas later subjected to severe storm or flooding to safer zones.

In the USA, some native American Indian tribes appear to have similar traditions. The Hopi in Arizona, for example, live atop the magnificent Grand Canyon and so see the spectacular work of nature first hand. But they believe in an earth spirit and, interestingly, that the circle is a key to communion with this sense of 'oneness'.

Hopi Indians were not immune to all the publicity about crop circles. Intriguingly when they saw the patterns that were formed in the fields thousands of miles away they claimed to recognize them instantly as a symbol of the earth spirit. They argued that they were a reaction against Western society's raping of the planetary eco-system.

Lines in the desert

The Nazca plains of Peru are vast and seem inhospitable to all bar a few hardy creatures but they were effectively farmed by the local Indians for a long time. In the period between about AD 600 and 1200 they seem to have created a wonderful pantheon of images on the desert floor.

Just as at Stonehenge these marks were formed in different eras across a long period of time, with certain kinds of patterns dominating each era. There is a wide variety. These include huge cleared areas, where the rocky floor of the plain has simply been swept away. Others are animal shapes or giant birds. But the most puzzling are the huge straight lines, rather like leys, that stretch many miles across the floor and often intersect or culminate in a cleared area. Some of the lines are also gigantic spirals that seem to wander for miles, going nowhere.

The lines themselves are quite easy to produce. Maria Reiche, who devoted much of her life to studying these mysterious patterns, found that it was a simple matter to expose the darker rock just below the surface layer in order to make the lines stand out spectacularly. There are limited opportunities for erosion in this desolate region and a line that is etched into the ground can easily remain for centuries.

The Nazca lines were not found until the advent of air traffic because they are largely invisible from ground level. At once there was intense speculation about the origin of the lines.

Since their full wonder is only appreciated from above there were those who suggested that aliens had made them – perhaps guiding the natives from spaceships hovering overhead. The founders of the 'ancient astronauts' theory, such as Eric von Daniken, even tried to show that some of the lines resembled runways and parking bays, although this is an illusion created only when you magnify small sections of some of the animal shapes (e.g. the claw of a bird) many times. In fact, any spaceship or aircraft landing on the tiny 'runways' that may perhaps be present would have had to carry very 'little green men' indeed.

Another idea was that the Nazca Indians might have had balloon technology. A dangerous experiment was tried to show that a device could be built from local products that would have been available a thousand years ago. The home-made balloon just about got off the ground but it was enough to argue that tethered balloons might have carried Nazca architects aloft so that they could supervise the work down below. In this theory, the fact that the symbols could be seen most effectively from the sky was meant as a tribute to the gods, to whom the lines may have been directed.

However, in 1992 David Browne, an archeologist with the Royal Commission for Ancient Monuments, reported on his research in Nazca and concluded that only 10,000 people working for a decade could have produced all the known lines. Their production has been spread over centuries of time and they cannot be regarded as an impossible undertaking which must have had some deep purpose for it to have even been contemplated.

Browne feels that the traditional interpretations, which follow the trend of astronomical calendars adopted for Stonehenge, do not hold scrutiny. Instead he found evidence that the lines in particular could be best thought of as not dissimilar to Devereux's 'Spirit paths'. Perhaps they were tracks that led people to huge cleared areas where religious ceremonies took place. As such, they were never really designed to be seen from the air.

It may be rather like our own creation of a motorway or freeway network. A traveller from beyond the Earth looking down on an old civilization may think this complex pattern would have involved huge amounts of time and money to create a pointless series of constructions best appreciated from above.

Awakening the dragon

The Chinese had a theory not at all unlike 'leys'. The Feng Shui (or dragon lines) system was said to mark out the spiritual essence of the Earth with key points at its intersections. A similar system is applied in microcosm to the human body and the nodal points are the locations where acupuncturists place needles to bleed energy or cure disease.

It is interesting to wonder whether the placing of stone circles and standing stones at ley focal points does not reflect exactly the same principle. If so, it is a remarkable example of parallel cultures developing identical ideas when it seems certain that they were fully independent of one another. Does this imply that there is some universal truth that was grasped intuitively by ancient man?

Attempts have been made to discover if there really is an 'energy' at these nodal points. The most consistent experiment has been the long-term 'Dragon Project' which has involved many hours of measurement at standing stones wired up to electrical, magnetic, ultrasound and radiation monitors. Pulses of energy have indeed been found in this way, particularly at sunrise. It is as if a charge is built up inside the stone and then released in a burst around this time of day.

The project was started in 1977 when Paul Devereux decided to bring together scientists, earth mystery researchers, folklorists, dowsers and psychics to see what each could contribute from their unique perspectives to our collective knowledge of these energies. There had already been anecdotal stories about people receiving 'electric shocks' by touching the stones or even of healing experiences occurring within the 'bath' of energies that they were said to present.

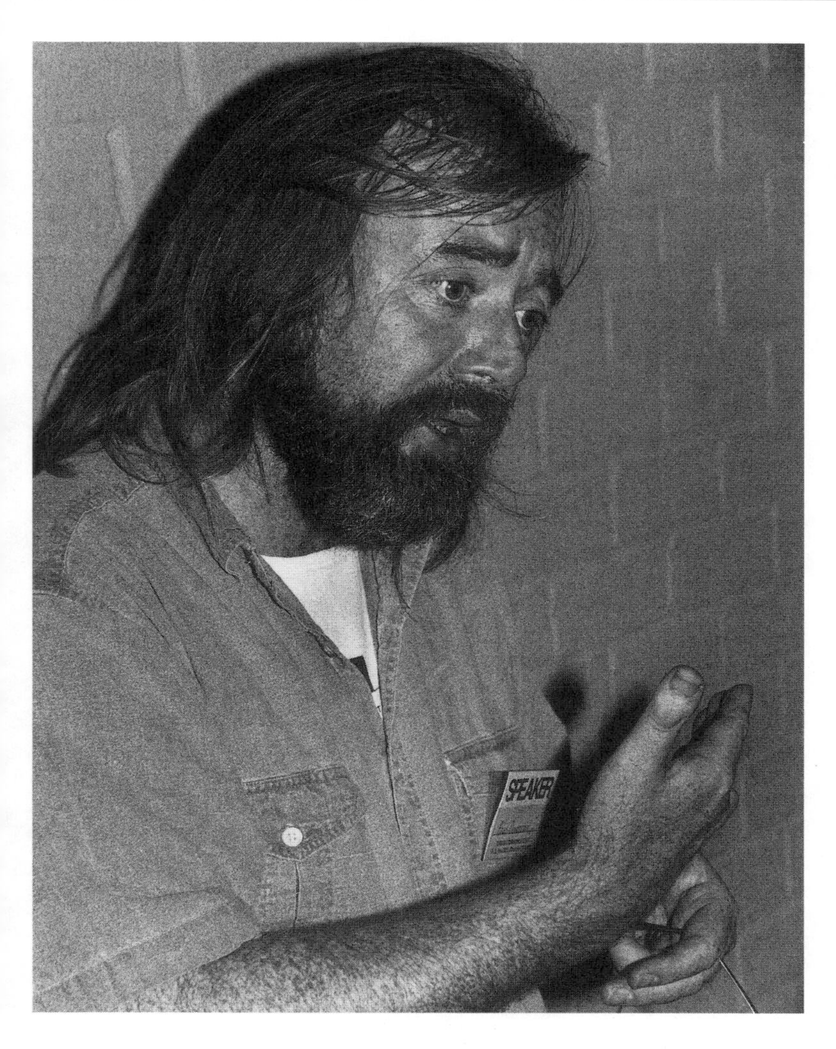

Noted earth-mystery researcher Paul Devereux who has pioneered much new thinking in the field and is a world-respected lecturer. He originated the Dragon Project to study earth energies from a very broad perspective

One of the leading protagonists of the project was chemist Don Robbins who measured ultrasound emissions at the Rollright Stones near Oxford in England. There was some evidence that the level of energy varied according to a number of factors, including the phase of the moon and cosmic radiation. However, the precise cause of the energy burst was difficult to determine.

Robbins at first thought that electrons within the rocks might have been excited by external forces creating a microwave emission. However, it was later speculated that the origin of the impulses might lie more in the surrounding rock, even in fault lines that run through the Earth.

Further experiments were of a very different nature, showing the diversity of procedure. Alan Cleaver, a specialist in Electronic Voice Phenomena (EVP) tried his equipment at Rollright. Essentially this involved a tape recorder shielded to prevent stray emissions. It is left to record the ambient silence and from time to time it has allegedly produced brief snatches of sound that appear to be voices. Unfortunately, these rarely make much sense and are usually so faint as to be almost impossible to hear. Their source has been variously ascribed to possible radio transmissions accidentally picked up, distant voices carried by the wind and certainly they owe a good deal to the imagination of the listener.

However, Cleaver reported that to his surprise he obtained some of the best results when working at Rollright. He recorded a voice saying ‘Raudiv’, apparently the commonly used

The Dragon Project in action at the Rollright Stones in Oxfordshire. Local co-ordinator Roy Cooper measures radiation field emissions with a geiger counter

nickname of Latvian psychologist Konstantin Raudive, one of the founders of the experimental protocol. The message was picked up in 1980, but Raudive had died in 1974. Among other messages from Rollright and elsewhere, were the remarkable 'Alan . . . This is Konstin Raudive' and 'Ghost is the listener', a phrase one can interpret in many interesting ways. Cleaver himself is insistent that when comparing recordings of Raudive's real voice and the one picked up at Rollright these are not the same. Who else or what else it might have been remains unresolved.

The latest experiments being conducted at circle sites involve a bold testing of consciousness as experienced through the dreams of subjects. Hardy volunteers are sleeping in burial mounds or at an approachable stone circle site and recording on tape all that they dream. These are then being centrally transcribed and comparisons made between the dreams that are emerging at each location.

It is hoped that some evidence may emerge that sensitive people can somehow 'tune in' to an essence of the location, possibly even 'tap in' to information that might have been imprinted into the stones and replayed in their minds as they slept. One possibility is the depiction of ancient rituals that occurred at this site. The experiment will probably report in 1994 and there is much anticipation with regards to what results might emerge.

An alien Stonehenge?

Some researchers report that it is not merely the Earth that has puzzling features. The planet Mars has them too. Of course, Mars has long been associated with science-fiction stories about dying races and alien invaders. The 'canali' (a word meaning channels or grooves but popularly mistranslated as 'canals') were recognized by astronomers a century ago. They provoked much speculation about possible waterways built by the Martians, and later, of an extraterrestrial equivalent of the ley system or Nazca lines.

Now we know that the canali were partly illusions due to the poor resolution of our telescopes and also dim perceptions of huge rents in the surface of the planet created by geological forces.

However, additional mysteries have recently surfaced which seem to have more substance and do appear to be connected with lines of power and focal points of energy, such as Stonehenge on Earth. Our spacecraft have now visited Mars several times taking orbital photographs, landing and collecting soil samples and filming the eerily coloured Martian sky and its bitterly cold and boulder-strewn surface. We have found no Martians, not even microscopic organisms. We have filmed no running water, although evidence is mounting that in the long distant past it may have existed there and some dried-up valleys look suspiciously like ancient river beds.

Mars has, however, recently incited perhaps the greatest mystery to emerge directly from three decades of spaceflight. Our orbiting missions have found a puzzling region called Cydonia where evidence of extraterrestrial life has been obtained. There are only a handful of photographs taken here in 1976 from which the 'Martian Monuments' theory has grown. NASA has not made any fuss about them and, when pressed, said that this was because they felt they were just an accident of light and shadow on the surface of the planet. This created an illusion of constructed objects. In truth these were ordinary rocks and mountains that merely look strange to the fallible human eye.

Such an option must be considered, but is made less tenable when we appreciate that the mysterious images were taken from different orbital positions with the sun at different angles. This reduces the likelihood of a light and shadow effect, while not eliminating it altogether.

Authorities

■ *Paul Devereux* is probably the key figure in modern earth mysteries research. He is editor of the house journal *The Ley Hunter* and has written many seminal works. In his recent important books that reassess the entire field, *Earth Memory* (Quantum 1991) introduces what he calls the 'holistic' approach to ancient sites, *Shamanism and the Mystery Lines* (Quantum 1992) discusses his new 'spirit path' and consciousness orientated approach, and *Symbolic Landscapes* (Gothic Press 1992) is a review of the current thinking on leys and stone-circle sites.

■ *Dr Terence Meaden* is the physicist and meteorologist noted for his research into crop circles. His radical look at stone circles suggests the deification of strange meteorological forces which were reflected by ancient peoples in their construction of stone-circle sites. This is outlined in *The Stonehenge Solution* (Souvenir, 1992)

■ *Don Robbins* is an inorganic chemist who took a PhD in magnetism and has been a mainstay in the Dragon Project's search for physical energy emissions from stone circles. His book *Circles of Silence* (Souvenir, 1985) describes some of the experiments carried out by the project.

■ *Alfred Watkins* is the antiquarian who really founded earth mysteries research. His book *The Old Straight Track* announces the rediscovery of 'leys' and was first published in 1925. It was reissued by Garnstone in 1970. Devereux (*see* above) and Nigel Pennick updated the concept of leys in their useful introduction *Lines on the Landscape* (Robert Hale, 1989)

■ *The Ley Hunter* is published from PO Box 92, Penzance, Cornwall TR18 2XL.

■ The Mars Mission publish their latest work in *Martian Horizons* at 31–10 Skytop Gardens, Parlin, NJ 08859, USA.

■ Stonehenge Viewpoint is an organization that collates research from a wide diversity of fields into stone circle and other ancient sites. It publishes regularly from 800 Palermo Drive, Santa Barbara, 93105, USA.

Power points

- There are ancient sites whose origin is unclear in almost every culture on Earth. They seem to involve a deification of forces from above and often include an awareness of hidden powers within the planet.
- Several cultures, including the Celtic, Chinese and South American Indians had concepts of, and created, straight lines linking many such sites. There are even hints that similar patterns may exist on the surface of the planet Mars. What ancient secret does this fact hide?
- Many religions have developed a view of the Earth as a living biological entity of which humanity is but an organism. The aboriginal culture of Australasia perceives a serpent winding through their dreamtime and channelling a life current. The North American Indians of Arizona claim to recognize modern symbols such as crop circles as signs of an angry planet. Is this an intuitive awareness of a belief now surfacing in western society as an ecological message?
- Scientific exploration at many sites, using ultrasound, microwave and other monitor equipment has found evidence of energies locked up by the rocks which can be released in sudden bursts. There are well-attested cases of tingling and 'vibrational' sensations picked up by people who have simply touched the rocks at ancient sites like Stonehenge or Avebury. Sunrise and sunset seem to mark peak times for such activity.
- Experiments are under way linking human consciousness, altered states and dream imagery with this power source. A modern-day equivalent of the 'dreamtime' is being rediscovered by so called 'earth mystery' students in a field with growing popularity, especially among the young.

In 1979 two computer specialists, Vincent DiPietro and Gregory Molenaar, digitally enhanced the surface pictures and confirmed that what the eye appeared to see was indeed quite real. At least if the eye was being deceived, then so was the computer.

In 1983 NASA consultant, Dr Richard Hoagland, was one of the first to take up the cause after seeing the significance of their work. Two photographs in particular taken by a Viking orbiter appear to show a straight-sided pyramid and, even more strikingly, a face with eyes, nose and mouth in human-like proportions. The rock from which this was sculptured was about a mile across. Hoagland attempted to find a mathematical relationship linking the angles of straight lines that pass through these sites, much akin to leys on Earth and similar to that which has been sought at Stonehenge. After much work, and to his astonishment there were countless clear patterns which seemed to be built around mathematical constants which imply an artificial, rather than natural, origin.

As time went by the evidence for both these lines and the 'monuments' continued to grow. Graphical analysis was tried and seemed to support the existence of a strangeness behind the face. More sophisticated computer-enhancement techniques became available. In previous examples this higher level of definition tended to put images that were mere illusions sufficiently over the threshold to destroy their artificial appearance. But not in the Mars images. What seemed to be details like a pupil within the eyes were uncovered, making the suggestion of intelligent construction all the more plausible.

Perhaps the most intriguing work of all was a three-dimensional computer simulation of the surface area eventually formed by combining all this data. This gives an undeniable image of a slightly eroded pyramid next to a human face gouged from rock and staring into space. The similarity to the pyramid and sphinx in Egypt is indisputable.

This mounting evidence won new converts. Dr Brian O'Leary is a physicist who trained with NASA as an astronaut planning to go on their manned missions to Mars before they were postponed. He told us that he was at first a sceptic but when he looked at the evidence he became persuaded that there really was something odd on the Cydonia plain.

But why make this object look like a human face? Is it not stretching credulity too far to think that Martians would look like us? The proponents of this evidence suggest reasons for this riddle.

Perhaps, they argue, the monument was deliberately put there by an extraterrestrial civilization just visiting our solar system long ago. They would have seen primitive humans on Earth and perhaps created this in our image so that it would be instantly recognizable by ourselves. We would only see it when our technology developed to a level that enabled us to reach Mars, because it is too small to be visible from telescopes on Earth. As such we would feel the urge to investigate further.

It is a romantic notion, but one that Hoagland and others are keen to try to prove or disprove. Interestingly, the theory sounds similar to 'The Sentinel', a short story by Arthur C. Clarke which formed the basis for the film *2001 – A Space Odyssey*. During 1992 Hoagland made pleas through the United Nations to mount an international exploration to Mars. His team ('Mars Mission') has also endeavoured to convince NASA to re-photograph the area with far more sophisticated cameras.

In August 1993, the new Mars Observer craft prepared for 'parking orbit' around the Red Planet after a journey of one year. As the electronics came to life, suddenly all communications with Earth were lost.

Some commentators speculated that either Martians had damaged the craft to stop it transmitting new pictures, or NASA already had the evidence and had faked the loss of communication as part of a cover-up.

In truth, this was the most dangerous part of the mission – a time when things would be likely to go wrong. But it leaves a lot of people wondering.

Fresh attempts to map Mars have been planned by both the Americans and the Russians.

Any pictures that are returned by the new 'Mars Observer' craft would be clear enough to solve the mystery as to whether this really is an alien version of Stonehenge or just another testament to human gullibility and our desire not to be alone.

FIELDS OF FORCE

The world awash

The Earth is a living planet constantly surrounded by electrical fields. The two polar regions are like giant magnets which drive unseen currents around the globe. Our upper atmosphere is bombarded by charged particles from the solar system and beyond, creating mysterious glows that light the night sky such as the aurora borealis or northern lights.

Most of us are only dimly aware of this electrical power that is seething around us because it is largely invisible. But it is possible to occasionally sense it. For example, many people report that they know when a thunderstorm is approaching. This is not some prescient awareness. It appears to be a sensitivity to the electrical fields crackling within the atmosphere and the static charges that are then building up. Acting rather like a lightning detector the body picks up on this and responds with physical sensations. These are often described as tingling skin, oppressiveness, tightening around the forehead and migraines. They are all a reaction to the power of the planet that we live on.

However, it is not only the biosphere of the Earth which contains these charged fields. They circulate within human beings and we all have a certain amount of body electricity and can store static charge just like a capacitor. It is easy to demonstrate this; many children love the game of rubbing a balloon against silk clothing and then watching how it will cling to their hand even when this is held upside down. Other more daring souls can be connected to van de Graff generators which produce huge amounts of harmless charge and store enormous levels of static in their bodies akin to those found within thunderclouds. This is enough to make one's hair stick out like a stiff brush. But there are a few people who report far more mysterious electrical phenomena.

My kitchen just blew up

Most people are familiar with domestic electricity fed through the national grid from generating stations to circuits within homes and public buildings. We may not understand how it works but we do take it for granted and expect it to be available. What we certainly do not expect is to find it wreaking havoc and leading to widespread destruction.

At Piddlehinton in Dorset in May 1990 things certainly did go awry. The local electricity board was inundated with complaints and protests from various residents when all at once their domestic appliances began to 'apply' themselves on their own.

Kettles clicked on and overheated. Light bulbs exploded. Video recorders switched on or off all by themselves. Television sets began to emit weird noises. As one spokesman for the mystified electricity board admitted, 'It was like something out of a horror movie.'

The blame was eventually put down to a freak power surge within the system. It was similar to the power surge which had hit a relay circuit near Syracuse, USA, in November 1965 and triggered a massive series of breakers (set to cut off automatically following any overload) placed within the grid system connected to Niagara Falls. As a result huge areas of New York State and north-eastern USA were blacked out with ensuing chaos. The precise cause was undetermined but some eyewitnesses reported seeing a mystery ball of light near the affected relay site.

This must be no comfort to someone like Frank Pattermore from Somerton in Somerset. Since 1982 his stone cottage has been similar to a war zone and nobody seems to know why. Lightbulbs and fuses blow with such regularity it is almost pointless refitting them. After television sets kept exploding and flames emerged from the ceiling the experts were called in. The entire house was rewired several times but nothing cured its phantom power mania.

The electricity board has spent over £50,000 in an attempt to cure the 'haunted' cottage but to no avail. They know that up to 2000 volts seems to be seething through the house but they do not know why.

The prime minister's office, responding to a plea for help from the disgruntled householder, found that their experts failed to solve the riddle. However, a physicist who researched the house measured heavy electrical interference in one part of the room. Even when everything electrical within the building was turned off the meter still measured power being consumed. There are those who think that a nearby telecommunications and a Ministry of Defence site might be responsible. But this has been denied. Meanwhile the power surges go on unabated.

Electric people

It appears that there are human beings who themselves generate so much energy they become a sort of biological dynamo. It has become an accepted problem of office life that some people just cannot work computer terminals. Whenever they go near them the screen plays unusual tricks and data can be lost.

These 'office jinxes' sometimes have an unfortunate association with the supernatural and they suffer as a consequence. Indeed it is interesting to note the obvious parallels with poltergeist activity. In the age-old mystery of the supernatural all manner of strange things reportedly occur, but in recent times electrical disturbances and exploding lightbulbs have been common. The tendency has been to associate these with mental anguish suffered by a 'psychic' victim at a specific point in their lives, but there now seems every likelihood that these are merely further instances of body electricity. If someone has a huge charge built up inside them it would be capable of scrambling a VDU screen on a computer or surging power through a lightbulb.

Jacqueline Priestman from Cheshire was one such person. Electrical equipment seemed to go haywire in her presence.

She would only have to pass close by the television set and it would switch channels. When she put a plug into a wall socket sparks would leap out. Once she held an iron and flames poured from it. The power was leaking out of her body into the coils of the system causing them to overheat.

Jacqueline was studied by Dr Michael Shallis, a physical sciences lecturer at Oxford University. He was able to establish that she was one of the rare group of people who had about ten times the normal level of electricity inside their bodies. But what could be done to help her? This was far from obvious. As it turned out she helped herself, without realizing it.

Mrs Priestman changed her diet and began to eat a lot of green vegetables. For some reason this seems to have defused the energy in her body down to more normal and manageable levels.

In another well-attested case of this type in April 1934 at Pirano in Italy, a woman was studied by a whole team of doctors from the University of Padua. They were particularly fascinated because at night her body literally glowed with a blue flicker of electricity visible through her skin. They noted that when these brief spells occurred the woman was asleep and her breathing rate and heartbeat had doubled. But the phenomenon ended as quickly as it

'Car stop' cases are remarkably consistent phenomena. Some strange electrified force field seems to cause vehicle engines and lighting systems to fail, as depicted here by Roy Sandbach

had begun before anybody figured out what was causing it. Interestingly, the woman was devoutly religious and had just engaged in an extended fast when the phenomenon started, suggesting a link with her diet.

May the force be with you

Another puzzling electrical effect is the so-called 'car stop' or cases of 'vehicle interference'. It is often thought to be a part of the UFO mystery (*see* page 52), but that is rather sidestepping the issue.

An interesting example occurred in November 1967 in the early hours of the morning at Sopley near Bournemouth. The spot was adjacent to a then secret radar research site.

Two vehicles were approaching a crossroads by the River Avon. One was a truck driven by Carl Farlow and the other a car in which a vet was returning from a late-night call along with a female passenger. Ahead of them the occupants of both vehicles observed a small reddish oval shape drifting directly across the road. They were bemused by it but became even more puzzled by the effects that it generated. Both vehicles lost power in their headlights. The engine in the car also lost power and eventually ceased to function altogether. The truck, however, kept going. The driver edged to a halt but the engine continued to tick over.

Eventually the small oval mass drifted away and the occupants of the two vehicles got out to swap notes. The police were called and the Ministry of Defence also sent investigators out to further enquiries. Mr Farlow's truck had to be started with a tow because the battery was drained. But its engine was still operating. The car battery was drained as well but its engine worked perfectly again after the strange floating mass had disappeared.

The only difference between the car and truck was that the truck was powered by diesel. There are several other similar cases where diesel-powered vehicles continue to operate when normal petrol driven vehicles (which use spark plugs) fail.

The vehicle interference effect, whatever causes it, is undoubtedly one of the most consistent scientific anomalies in modern times. Over five hundred detailed investigations have been carried out during the last forty years and a wealth of data exists. No answers have been found. But it would seem likely that a free floating electrical energy field is the cause and that these occur in certain locations at certain times for reasons which are not yet apparent.

In other examples of this strange phenomenon a motorcyclist in Barnard Castle, County Durham found that the infra-red heat generated by the floating mass was enough to evaporate all the water from his machine on what was a very rainsoaked night. An American police patrol car was forced off the road in Nebraska by an oval mass that seemed to be sucking power (or gaining life) from the overhead power lines nearby. And in one extraordinary case from Tasmania, the energy allegedly pulled the vehicle along the roadway as if it were magnetized.

Human beings involved in these strange affairs frequently report physical effects. These commonly include watering eyes, tingling skins, hair standing on end as if attracted by static forces and sometimes what appear to be minor responses caused by exposure to infra-red, ultra-violet or X-ray radiation. These again strongly infer the presence of radiating energy fields surrounding the glowing mass.

Unfortunately, the popular idea is that these forms (most likely to be natural atmospheric energy fields) are instead what we term UFOs and, by association, alien craft piloted by little green men. This has held back too many scientists from studying them carefully. When they do we may learn something about the energies that encompass our planet.

The way home

For many years scientists have puzzled over the remarkable homing instincts of some animals. Birds fly in unison towards a home in the winter sunshine. Millions of butterflies leave their habitat in northern USA and converge on one small area several thousand miles south with regular and unerring accuracy. Just how do they achieve this miracle of navigation?

Racing pigeons have the ability to find their way back to their own roost and competitions are held across hundreds and thousands of miles to see who can do this in the fastest time. But occasionally, and for no obvious reason, this instinct suddenly deserts them.

In one competition in 1978 hundreds of pigeons released from Scotland never reached their destination in England. The entire flock vanished into the blue. There was speculation about abduction by aliens but the rather more likely explanation seems to be that something scrambled their inbuilt homing sense.

Magnetite (magnetic iron oxide), as used to form a needle in a compass, acts like a magnetic pole and is found inside the skull of pigeons. By using this almost like a radio beacon they can attune to the lines of electro-magnetic energy that cross the Earth's surface and so find their way to wherever they want to go.

However, at times there can be a disruption to this free flow of energy. Intense solar flare activity on the sun's surface can send a bombardment of cosmic particles rushing to Earth at the speed of light. Eight minutes after leaving the sun this invisible and (to us) harmless wave of energy hits the atmosphere. Here it can seriously distort the ionosphere, an upper layer of the air which we use to bounce radio waves back to Earth for communication. For days on end we can have problems with our transmissions because of this interference and, it seems, the homing instincts of animals such as racing pigeons are equally vulnerable to this silent invasion from outer space.

In experiments a surprising variety of life-forms have been found to possess rudiments of this magnetic sense. Everything from termites to sharks appear to use some inbuilt magnet to assist in their directional capabilities. It has even been reported that bacteria in a laboratory show a propensity for aligning with the magnetic field of the Earth.

Experiments have again demonstrated that by encasing the animals in an intense electro-magnetic field of artificial origin this totally disorientates the creature.

Probably the most amazing example of homing instinct is that associated with cats. There has long been a tradition of folklore that you can never desert your cat. If it wants to it will always find its way back to you. So much so that pet owners know that as a matter of routine if they move house they have to take significant precautions to ensure the cat stays put until it realizes this is its new home.

In 1992 researcher Paul Sieveking drew up a league table of what he called 'pussy trekking' from all the available sources and found some astonishing examples. These cases come from all over the world and range from the relatively mundane to the almost unbelievable.

Beatle, a two-month-old kitten, given to a boy in his London flat in late 1986 ran off and five

Animals, such as pigeons, have a remarkable ability to navigate and fly in formation by attuning to the electromagnetic energy fields generated within the earth's atmosphere

months later returned to its original home in a Kent farmhouse after travelling 75 miles through snow and ice.

In another case, Sam had moved with his owners from Wisconsin to Tucson, Arizona, USA, in 1986. When they returned to Wisconsin in 1987 he was left to be adopted in his new state. Yet, four years later, in 1991 his owner chanced to visit the new occupants of her old house in Wisconsin and there was Sam, living rough in the garage. He had been there for some time, it seemed, and nobody had realized that he had made this 1500-mile trip all by himself.

In Australia, Howie, a three-year-old cat, travelled across the dangerous bush country from Adelaide to the Gold Coast of Queensland, well over 1000 miles after being left behind when his owners moved. It took a year but upon arrival, with bleeding paws and covered in sores, he purred in delight to find his owners at their new home gaping open mouthed at this miracle return.

Human homers?

If so many animals have this uncanny instinct, is it possible that humans possess it as well? It is, in fact, thought very possible that we do but that it might be misinterpreted as ESP (extrasensory perception).

In one experiment human beings were led into the countryside by a psychologist and although blindfolded were asked to 'guess' the direction home. Using only intuitive instinct (or, if you prefer, simple guesswork) they were able to point the direction more accurately than a control group who were not blindfolded at all.

This seems to suggest that it is very much an

instinctive response and is notable only when it serves as the main source of information. The instinct is presumably located in the right hemisphere of the brain controlling intuitive functions. When all the more overt senses are working normally we may tend to suppress any information picked up in this subconscious way and rely more upon data being processed by the left hemisphere of the brain. This area is responsible for logical and rational thought and our culture is taught to regard that as supreme.

It would appear possible that what we refer to as 'a little voice' or simply a 'gut feeling' about something is, in fact, subconscious awareness of this less-trusted but highly accurate kind of information.

If this is received by the brain alongside other, seemingly contradictory, rational thought most of us will have become conditioned to ignore it or, at best, regard it as less important than more dominant logical thought patterns. As such we may be allowing what is a very natural instinct built into all of us, to atrophy in a world where such things are considered less meaningful.

There is ample evidence for this. In what we call 'primitive' cultures where Western technology has not yet arrived these abilities are accepted as normal. Australian aborigines often trek across large distances without maps or compasses and have been found well capable of 'tuning in' to the presence of others in the same tribe. Similarly there seems an awareness of danger which we might dismiss under pseudo-scientific names such as a 'premonition' but it is often little more than a capability to attend inward. This basic instinct may well be something that humans have always had but which today our society has discarded, perhaps to our disadvantage.

The wife of round-the-world yachtsman Chay Blyth reported that she was in an English restaurant eating while her husband was thousands of miles away sailing a catamaran. At the exact moment this had overturned and Chay was trapped in icy seas for many hours, Mrs Blyth was hit by a wave of emotion. She just 'knew' something was wrong. There was no premonition, just a presentient awareness.

The Princess of Wales, according to Andrew Morton's version of events in her life, seems to have had a similar subconscious awareness of danger when her husband, Prince Charles, was involved in a fatal avalanche on the Alpine ski slopes from which he luckily escaped. This is not anything we should term strange or supernatural. It appears to be so commonplace and widespread that it is a fact of life. Only our scientifically dominated society, which has as yet failed to completely understand how it works, finds it difficult to accommodate. No other culture regards it as a problem.

People whom we deridingly term as 'psychics', and dowsers, using a forked twig to locate water and minerals, may merely be those who have learnt to use a subconscious sense we all possess. Most of us push this so far beneath the surface that it rarely emerges. But, if so, are we not the poorer for it?

The humadruzz

In the late 1950s people in Britain began to report a most unusual noise. It was like a combination of a humming, droning and buzzing sound. It was particularly prevalent during the middle of the night and at first was written off as mere imagination.

By 1962 the term 'humadruzz' had been invented to illustrate its character and enough reports were being made for it not to be doubted any more by some scientists. However, nobody knew what it was.

A government enquiry was eventually launched by Lord Hailsham. Victims of the persistent hum were placed in specially soundproofed rooms. Some still heard the noise. This seemed to destroy the original theory that it was simply the buzzing of faulty electrical circuits in the bedroom or some other ambient noise such as a local sub-station.

Eventually the investigation ended with the suggestion that the source was most likely to be person-centred, perhaps medical since others in the room at the same time often said they could hear nothing. The whole thing was quietly forgotten as a temporary British aberration.

But it was not temporary. Nor was it confined to Britain. The hum was being heard all over the developed world. In late 1991 it struck the community of Hueytown near Birmingham, Alabama, USA. It was compared to a dentist's drill by some of those who heard it – a grinding, penetrating hum that bored into the skull. As this is a mining district, the blame was first laid upon underground activities, much as it had been in Britain during some of the cases in the 1960s. Here the idea grew into a localized legend that a race from some subterranean land was burrowing beneath the Earth when people were in bed.

One researcher who probed the Hueytown noise said that this sort of thing occasionally happened when freak combinations of atmospheric conditions and local geography focused noise from miles away like an echo chamber. However, no obvious source for this particular Alabama sound could be traced. The mines had installed no new equipment in recent years and yet still the humadruzz went on.

In this instance the Hueytown case was understandably treated by its local victims as unique. There was no apparent awareness that this is a global problem of more than thirty years duration – although that fact inevitably effects any attempt at an explanation. Indeed at the same time as Hueytown was humming the inhabitants of the rural areas surrounding Stroud in Gloucestershire, England, were the latest to hear the mystery noise.

These events concentrated mainly in December 1991 and January 1992. Many local people blamed unusual activity at the nearby GCHQ 'spy' and telecommunications establishment in Cheltenham but this was strongly refuted. Eventually a government environmental department investigation was mounted with a £50,000 budget. It failed to come up with a universally acceptable solution. However, the researchers did at least lay to rest another charge that was often made about the humadruzz: that those who hear it suffer from tinnitus, a problem which generates a ringing noise inside the head.

The Department of the Environment, in justifying the research, claimed that it had only just been alerted to the problem but was already getting five hundred complaints a year in Britain alone. Evidently they were not familiar with the previous government study of the sound carried out almost three decades before.

The noise has only been tape-recorded once – at Hough Green in Cheshire. By moving the tape recorder from the bedroom and holding the microphone out of the window it was demonstrated to originate outside the house. In one case during the day when the hum manifested itself in 1973, a youth rode his bicycle off the road as a result, he said, of the penetrating noise. But still no solution was found.

The recorded noise faded in and out, like a distant sound blown on the wind, and had acoustic characteristics of a grinding, drilling and whining. Some speculation was offered that it was late night repair work on the nearby railway track but British Rail denied this suggestion. Again the noise disappeared without resolution.

The humadruzz is not simply an academic problem according to a report by acoustic experts. Some of the persistent victims have had to take tranquilizers in order to sleep. In one case a young child became violent as a result of the noise and attacked other people and herself, and there is also an alleged suicide precipitated by the noise.

So what is the cause? A popular theory is that the sound is created by water forced at pressure

THE H-U-M

HELP ME TO ESCAPE

FRESH evidence about that mystery HUM—reported in last Wednesday's Daily Sketch—is now flooding into the office.

Each post brings hundreds of new facts from readers. These are being carefully sifted.

We described the nerve-wearing effects of this noise—estimated to be pitched between 14 and 20 cycles—which only a minority can hear. You emphasise that even to this minority THE HUM is a nation-wide menace.

Here is a selection from your letters:

LESLIE WATKINS takes over Pungent Post with your letters on The Hum

STRAIN

NIGHTLY

DEPRESSED

POWER

VICTIM

SPOT ON

NERVES

THE HUM

The Daily Sketch tests a victim in the BBC's sound-proof room

Daily Sketch 21-9-60

Mrs. Zameczník submits to the Daily Sketch test to track down THE HUM.

By Leslie Watkins

CAN THE HUM—the mystery noise which is baffling thousands in Britain — penetrate a sound-proof room? YES, it can.

READERS SAY...

The mystery hum is a fascinating phenomenon reported by many people all over the world. Here, the *Daily Sketch* of 1960 reports on experiments to try to explain it. It is still unidentified

Authorities

■ *Dennis Bardens* is a journalist who has collated many stories of unusual animal behaviour which relate to things such as their homing instinct. His book *Psychic Animals* (Robert Hale, 1988) is a rich store of patchily documented but intriguing anecdotal accounts.

■ *Dr Michael Persinger* continues to publish scientific papers updating his research, but his invaluable opus work was *Space-Time Transients and Anomalous Phenomena* (Prentice-Hall, 1977). He has done more than anyone to establish a link between atmospheric energy fields, strange phenomena and physical effects.

■ *Dr Helmut Tributsch* is a specialist in the field of animal disturbance related to possible electrical and chemical interference. His book *When the Snakes Awake* (MIT Press, 1982) is an excellent study of the way in which many forms of life respond to environmental changes. It is especially useful to see how in some cultures, such as the Chinese, the abnormal responses of animals are used as effective predictors of imminent seismic activity. Tributsch looks at how they may detect chemicals and fields surrounding them and note changes in these as the Earth starts to move – thus giving them an advantage over more 'blinkered' humans.

■ *Dr Lyall Watson* is a noted zoologist and researcher into strange phenomena. He has toured the world investigating unusual happenings. In his book *The Nature of Things* (Hodder and Stoughton, 1990) he has taken particular note of the interaction between human beings and electrical fields and equipment, reporting on latest research across a range of scientific disciplines which effect the view that mankind may indeed be a victim of the environment that he lives in.

■ *BUFORA and CUFOS* are UFO research associations: BUFORA is the British UFO Research Association and CUFOS is the J. Allen Hynek Center for UFO Studies based in Chicago, Illinois, USA. In 1979 both issued independent reports on the vehicle interference effect which stand as valuable dossiers on the phenomenon. These books, edited by Geoff Falla (BUFORA) and Mark Rodeghier (CUFOS), provide a complete listing of all the cases known to that date – although many more have been added since. They also have some discussion by experts on what may cause the events to occur. Neither presumes that the cause must be alien contact.

■ BUFORA, Suite 1, 2C Leyton Rd, Harpenden, Herts AL5 2TL

■ CUFOS, 2457 West Peterson Ave, Chicago, IL 60659, USA

through underground pipes, but these have only been installed gradually, well after the onset of the problem. Some clue that originates in the mid- to late 1950s seems to be required as this period is the genesis of the stories.

The theory which appears to work the best is that the humadruzz is created by microwave energy. These transmissions began in earnest during the 1950s and a network of repeater towers now beams signals all round the world. At certain spots these beams interact and create interference patterns. It may well prove that hums focus in the same places.

Normally, this energy is invisible and undetectable by humans. But it is considered possible by some neurologists that in certain situations microwaves may be absorbed by the brain cortex and turned into sensory stimuli akin to humming noises. If so, this is an unexpected and growing danger resulting from our own technological advancement.

Illness from the sky

Dr Michael Persinger is a neurophysiologist from Laurentian University in Canada. He has specialized in research on how electromagnetic and other energy fields affect the human brain. As a result he has discovered alarming evidence that the humadruzz may only be a minor aspect of a much larger set of symptoms.

He has discovered by experiments in which relatively small microwave and other energy types are beamed at humans, that a change of state of consciousness can result. Not only are unusual auditory sensations recorded but visual ones as well. It is even possible that a form of blackout not unlike an epileptic seizure can briefly occur from very intense concentrations of energy, and at times can nausea and headaches.

Dr Persinger and his team have proposed what he calls 'transients' or small pockets of energy that float free in the atmosphere riding along magnetic fields. These can be created by a variety of geophysical processes and may often be quite invisible. They would normally only be detected by the effects that they generate when the brain absorbs and reacts to their energy.

This is fascinating not only in view of the humadruzz but because it has other repercussions. There have been several recent investigations into electromagnetic pollution – with sickness created by no obvious physical cause but as a result of radiation in our surroundings which originates from our rapidly expanding technology. Just as atomic bombs have deadly, but invisible, fall-out, so, it is said by some researchers, these energy fields may provoke physical ailments in those who by chance are exposed to abnormal concentrations.

Residents of Fishpond Bottom in Dorset felt that the newly laid high tension cables spanning their quiet village were somehow responsible for the spate of headaches, sickness and dizzy spells suffered by the villagers. Although an investigation by physicist Dr John Taylor seemed to offer some support, the official verdict was that there was no link.

In September 1991 a Pennsylvanian housewife tried to sue the local power company alleging that the high tension lines nearby made her disorientated and gave her headaches. They denied the claim but she then alleged that domestic electrical equipment was charged up even when not switched on and that her children regularly played with neon light tubes under the wires because the strong energy field in the area caused them to glow like the 'light sabres' in the movie *Star Wars*.

Indeed, the latter complaint is a common one from residents near power-line networks. In one case where ball lightning (a rare type of energy ball which can enter homes) flew directly underneath a neon tube at Changi, Singapore, it caused it to glow despite the power being switched off. From evidence such as this it seems that the field induces energy into the circuits which may then operate independent of the mains. If so, one is forced to wonder what

Forceful facts

■ The Earth is bathed by powerful forces such as electrical energies, magnetic fields and invisible cosmic radiation. We are still discovering the effects these forces have on living organisms.

■ Both the natural environment and the electrical fields within human bodies can be affected by these forces and this interaction may well be the cause of strange phenomena, for instance altered states of consciousness.

■ Overloading or interacting energy fields can cause the environment to audibly 'hum' and our mechanical systems, such as car engines and lights, to be mysteriously affected. These forces may even 'glow' by igniting gases and drifting through our atmosphere. In the past they were deified by the ancients as 'spooklights' and are still being deified today as extraterrestrial 'flying saucers'.

■ If these forces course through people's bodies they can tingle, glow or emit powerful leakages which create malfunction and interference on modern equipment such as computers and telephones. These are frequently misconstrued as being signs that the person is 'possessed' by demons or that they have 'psychic powers'.

■ The interaction between disturbed energy fields inside a human body and the unbalanced energies within the natural environment can trigger physical effects such as undiagnosed illnesses and bodily disfunctions. These may be attributed to a paranormal cause. This wave of sickness is only gradually being recognized and termed 'electromagnetic pollution'. The only difference between this and poisoning from toxic waste is that you can see, and easily eliminate, toxic waste. Illness triggered by immersion within an atmosphere of seething energies can be hard to detect. It will usually be misunderstood as the reported ill effects have no physical cause at all and are thought to be psychosomatic.

Human beings can respond to electromagnetic radiation. Residents of Fishpond Bottom in Dorset allege severe effects from overhead powerlines. Can natural energy fields also produce physical illness?

happens to a human brain cortex that is similarly inducted with such powerful energy.

Electrical UFOs?

Although we cannot normally see these 'transients', Michael Persinger thinks that it is possible that they can become visible at times. In the case of the phantom energy causing neon tubes to glow the very strong localized fields excite the gases in the tube, which then glow through incandescence. It is quite feasible that in nature a similar form of induction might cause excitation of the atoms within pockets of natural gas. These might be floating around rivers or marsh land or any other suitable terrain during appropriate atmospheric conditions.

The result of such a process would evidently be an eerie floating mass of glowing energy, a bit like a giant neon light tube. The key difference might be that, being at the centre of a field of intense energy, anyone who got too close might suffer physical effects and any electrical equipment, such as a car battery might react.

It is certainly interesting that we do have many cases on record that appear to demonstrate Dr Persinger's propositions – such as huge oval masses that glow rich colours and can create physical effects in witnesses or even vehicle interference. The evidence is extraordinarily consistent, but has been rejected by most scientists on the grounds that it relates to UFOs – which, of course, from their viewpoint do not exist.

Were you to encounter such a giant floating neon tube and suffer physical illness or note your car reacting there seems little doubt that you would claim to have encountered a UFO. In the strictest definition, a UFO would be precisely what you did confront. The problem arises as to how our society would then perceive this phenomenon in alien terms.

When cases of what ufologists call UAP (unidentified atmospheric phenomena) are carefully studied some interesting patterns emerge. If there is a lot of water vapour present in the atmosphere, for example, when the sighting occurs near a river or in damp conditions, the floating mass tends to glow red. At other times it may be blue or white. This matches what we might expect for the various molecules that would be present in the atmosphere would glow with appropriate hues. This seems to offer further support that electromagnetic transients may indeed exist.

CROP CIRCLES

Symbols from space?

Most people have seen a crop circle. If they have not become one of the many tourists who have flocked to southern England during summer months in recent years they will have seen pictures of these strange markings on television news broadcasts or in newspapers. The circles have become the eighth wonder of the supernatural world.

Yet the mystery of these swirled patterns gouged intriguingly into cereal fields goes a lot deeper and is a good deal older than is generally realized. It also has a scientific aspect which challenges the popular view that this phenomenon must be supernatural in origin.

The circles do look artificial. There can be no doubting that. The crisp, sharp edges and the intricate swirls and layering found within their neatly pressed down centres all suggest to the romantic eye that no force of nature could have been responsible. The marks appear, usually overnight, in fields that were commonly well removed from prying eyes. As such they have captured the imagination of millions; from scholars to mystics and UFO buffs to artists seeking inspiration, they became a reflection of the strangeness of our planet.

Of course, the contention was that no humans could have made them. But somebody – or something – clearly did. As they are far better seen from the air, rather like the lines on the Nazca plains, it was an inevitable and almost immediate speculation that their origin was extraterrestrial. Perhaps this was the long dreamt of contact from another civilization; our first real proof that mankind was not alone in the cosmos.

Before long supposition had become proposition and speculation was accepted fact. The race was on to decode the meaning of these symbols and figure out just who was in contact and telling us precisely what.

A typical pair of crop circles swirled into a field at Westbury, Wiltshire in 1988. Sceptics believe that hoaxers use the vehicle tracks to enter the field although this is not always obviously the case

Crop circles: The official story

The very first mention of a crop circle came in the local press in Wiltshire in August 1980. Very few people noticed this small item. Farmer, John Scull, had discovered a circle some 60 feet (18 metres) in diameter in one of his oat fields near Westbury during May of that year. Assuming it was a freak of the weather he ignored it, then ploughed it over. But some weeks later two more appeared in nearby fields. These were also big, with almost perfect symmetry and circular crop rims standing tall like soldiers on parade. But inside, the stalks were swept to the ground as if hit by a blast of compressed air. The crop was still growing. It had not been damaged, just laid over gently with strange precision.

The fields around the circles were undisturbed. There were no tractor lines in these fields. If a hoaxer had attempted to get inside the fields to create these marks, they would have left tell-tale tracks, just as the farmer did when he first stepped in to observe these puzzles at close hand.

Intrigued, but not especially concerned, John Scull decided to report the matter to the local press to see if anyone had an explanation. A team of UFO experts led by Ian Mrzyglod from Bristol and a physicist turned meteorologist, running a private consultancy on wind damage, were the only people concerned enough to take a look before the fields were harvested. Later they got together to discuss what might have happened, conducted an excellent investigation and filed sober and very rational reports with their respective organizations.

However, other more vociferous UFO enthusiasts were not so restrained. They seized upon the story to suggest that a giant spaceship must have landed in Wiltshire. They were disappointed when few people took them seriously, the national media showed no real interest and further circles did not appear during the coming autumn, winter and spring.

Over the next three years a few circles formed from time to time, but only in the summer and mostly in Hampshire and Wiltshire. There were brief flurries of press interest, especially in July 1983 when several appeared. These had a very odd pattern – comprising one big circle surrounded by four smaller ones like satellites. The media interest was very short-lived.

However, by 1986 two UFO writers, Pat Delgado and Colin Andrews, had begun filing regular reports with a journal called *Flying Saucer Review* (FSR). This famous magazine for UFO buffs was a staunch endorser of alien contact claims and many of its avid readers were inevitably excited by the thought that these marks could be a significant new step towards contact by some unknown intelligence. The ideas soon began to flow thick and fast, bringing into the debate all manner of odd events, such as a tragic air accident that occurred not far from where a circle had appeared months before. But was this mere coincidence or was there a more sinister reason?

As if in response the circles were becoming more numerous (from ten to fifty or so each summer) and changed format every year – for instance adding rings around the circles but largely sticking to a simple design. Single circles still represented 90 per cent of all patterns found by now regular aircraft flights and keen observers who were scouring the fields of Wessex during the warm summer weeks.

In July 1989, with the circle season now as predictable as any calendar, Andrews and Delgado left their FSR heritage and brought out a book called *Circular Evidence*, published by the Bloomsbury Press. It had many colour photographs, including spectacular ones taken from the air by amateur pilot 'Busty' Taylor. It was visually appealing and attracted widespread press attention. Immediately in its wake came far more circles than ever before, formed in even more spectacular designs.

By 1990 the circle mystery had spread around the world. Andrews and Delgado's book had become an international best-seller, other titles were appearing in the new genre, crop circle journals and research groups had been launched and the term 'cereologist' applied by the experts to their subject. As news and pictures reached the rest of the world so circles started to be reported far and wide. No longer was this an exclusively British phenomenon. However, the best and most graphic circles were still to be found in the fields of Wessex.

The years 1991 and 1992 were golden days for circle researchers. Television crews from all around the world flooded in to film the latest patterns and to interview celebrity cereologists. Monitor exercises were set up at sites where circles regularly appeared, hoping to catch one on film as it was manifesting. Sadly the circles failed to appear when the camera teams set up.

As for the markings themselves, they became almost ludicrously excessive. No longer were they merely circles and rings. Now there were straight lines, curls, swathes and huge formations of assorted geometric shapes (quickly dubbed 'pictograms'). Then, when even these seemed to bore the press photographers and tourists, all sorts of animals from whales (1991) to snails (1992) began to pepper the fields.

Cereology was split into three warring camps, often fighting bitter verbal battles with one another. Some clung to the view that the circles were alien messages and desperately sought to translate the symbols. Others said that this was all too ridiculous. The complex patterns clearly were the result of an intelligence, but one that was very much earth-based and had a love of practical jokes. The third group said that while hoaxing was rife there were real circles formed by natural energy forces not yet fully understood. Scientists from universities all over the world, especially Japan and the USA, joined the observers to carry out serious experiments.

Then, in September 1991, many cereologists were devastated by the claims of two retired artists, Doug Bower and Dave Chorley, who said they had created hundreds of the patterns since 1980 – including all of the pictograms. They had even signed the last twelve or so before they owned up to the scam. Sure enough, the pictures of what many sources had proclaimed 'real' circles showed their 'D' and 'D' signatures.

For many that was the end. Crop circles were no more than a giant con trick. Yet a hundred more still formed in 1992 when Doug and Dave insist that they had retired and made none.

Gradually the researchers began to take up their studies again, some alleging that a disinformation campaign was seeking to destroy the subject, and that key figures inside the research community, principally those who opposed 'alien' theories, were government agents. Bizarre conspiracy theories became dominant.

Still the circles continued to appear.

Crop circles: The real story

Once the phenomenon was given a name, more cautious researchers realized something very important. Either these circles were the result of an intelligence (be it human or non-human), in which case they may have suddenly appeared as they had indeed seemed to, or else they were the result of a natural phenomenon which had never been recognized previously. If the latter case were true then evidence should exist to show that circles have always formed but nobody had collated any reports until the media took interest.

Serious investigation, involving an intriguing liaison between the British UFO Research Association (BUFORA) and a meteorological research team called TORRO (Tornado and Storm Research Organization) began to look for such data. In 1986 BUFORA research officer Paul Fuller and Jenny Randles, director of investigations published what was the first-ever booklet on the subject called *Mystery of the Circles*. In its subsequent expanded format as *Controversy of the Circles*, this became the UFO group's biggest ever selling publication. In 1990 they also wrote *Crop Circles: A Mystery Solved* (Robert Hale) which sought to show that there was no alien contact and a lot of hype. The circles, they alleged, were the result of two things: hoaxing and a natural phenomenon created by rotating air currents and electric fields.

The BUFORA research team was able to bolster this claim by pointing to the rapidly growing evidence that circles were not a new phenomenon. They not only clearly pre-dated the first hoaxes made by Doug and Dave (in 1976) but some even pre-dated the twentieth century. Many farmers described circles that had appeared on their land in their youth, or tales handed down from earlier generations. They had assumed these marks to be the result of freak weather and not reported them beyond the immediate locality as they caused no damage.

Scientific publications were scoured and some examples of circles were found perhaps dating back almost a century. They were treated as a curious anecdote or scientific anomaly in the magazines of the day.

There were even some folk tales which seemed to refer to circles. One concerned a field in Hertfordshire said to have been mowed flat overnight in August 1678 with legend attributing this to the devil. The earliest circles yet found come from Holland in the sixteenth century.

Fuller, Jenny Randles and others also uncovered numerous eyewitnesses who professed to have seen circles form. Their stories were consistent and seemed to refer to an invisible rotational air mass descending from above and touching the ground for a brief moment before disintegrating. The very short life of the air mass could explain why the circles are so precise.

All the cases of historical circles (totalling almost 150 by early 1993) and the eyewitness observations (approaching forty) refer to single, very simple circles. Not one describes anything as complex as the patterns, pictograms or animal shapes that have come in the wake of the widespread media attention since *Circular Evidence* was published in 1989.

These findings were not popular. Indeed they were widely discredited by both cereologists and UFO researchers who considered them heretical. The more rational investigators, which included scientists such as Dr Terence Meaden who had pioneered much of the theory behind the research, were accused of selective logic and finding examples of wind effects that only resembled circles but were not identical to present-day patterns. However, in some cases the similarities to contemporary accounts are too exact to be doubted although these are still to be established.

The probability is that the simpler type of crop circles have occurred for centuries. This does not prove that crop circles are not produced by alien contact but it does make both this option and the claim that they are all the result of modern hoaxing much less probable. The recurrence of these circles over the centuries strongly supports the view that they are the result of a naturally occurring phenomenon that has simply been unrecognized before. However, while the circles continue to capture the imagination this view is one that remains controversial and is unlikely to be accepted fully.

Since 1989 many fantastic shapes have appeared in fields, such as this geometric pattern at Barbury Castle near Swindon in July 1991. Most, perhaps all of them, are hoaxes rather than alien symbols

Worldwide circles

The crop circle phenomenon may have come to our notice in southern England but it seems to be a global problem. Indeed, Britain is no more blessed than several other countries if one considers the simple circles traced back over hundreds of years, and not the more dubious pictograms or animal glyphs that have only appeared in recent summers.

In Australia crop circles were recognised as a major mystery well before the events in Wiltshire in 1980. In January 1966 at a reed-filled bushland swamp at Euramo near Tully in northern Queensland perhaps the first significant incident took place. The hoaxers, Doug and Dave, claim this case inspired them but the idea lay dormant for a decade after one of them moved from Australia to England.

Early one morning George Pedley, owner of a neighbouring farm, came into Horseshoe Lagoon, which is part of a sugar-cane plantation south of Cairns. Ahead of him was a dark oval shape rising from the water. It climbed into the air sucking a mat of reeds upward and these were swirled into what we would now recognize as a single crop circle. It was photographed and had the same precise spiral centre as the others. The reeds floated on the surface.

Investigators from UFO groups quickly proved that a hoax was very unlikely in this case. Unlike the calm of an English corn field this bush had crocodiles and the lagoon was infested by deadly taipan snakes. A hoaxer foolish enough to create the reed circle in 1966, or the many others in the same area that have been reported since, would risk more than prosecution for such efforts.

Jenny Randles visited the Queensland bush in 1991, ironically the very day that Doug and Dave told their story to the press. The Tully river area is full of circle legends and local aborigines have traditions dating far back describing strange orange lights floating around. The first-ever crop circle monitor was set up in the lagoon during January 1968 and an automatic camera was triggered one night when a circle formed. Sadly the film disappeared on its way by post to the laboratory.

Samples from the Tully reed circles were checked for radiation and other chemical changes. Nothing was discovered, although the crop itself quickly died.

Today an even more active region for crop-circle formation is in the Mallee wheat belt on the Victoria and South Australian border. During the summers of 1972 and 1973 quite a number of small circles appeared at Bordertown and Speed. Sceptics suggested that they were caused by kangaroos nesting. In a case at Wokurna meteorologists proposed a very similar explanation to that offered by Dr Terence Meaden for recent British circles – that of a powerful vortex.

In December 1988 the Jollys, farming the Mallee region of Victoria, discovered many small circles on their land. During this time the sky glowed mysteriously above the fields and a vortex of energy was seen rising upward into the air from the spot where some circles were later uncovered. An aerial search traced no fewer than ninety circles – something cereologists today call 'grapeshot'. These look like raindrop splashes and can cover a large area of field. They are each usually only a foot or so in diameter.

Australia is by no means the only country where crop circles have been recorded. They are fairly common in Canada and New Zealand, both of which have documented cases preceding the first hoaxes by Doug and Dave. In 1991 and 1992 Canada seems to have better examples of simple circles than Britain. Summer 1992 was especially poor in Britain – a fact that some meteorologists believe may have been connected with the relatively bad weather.

In 1993 no fewer than forty countries seem to have records of impressive genuine crop circles. However, while there are some examples from the USA there are fewer of them than we might expect, given the huge wheat belts that cover the country.

Physicists such as Dr Terence Meaden have noted that the American wheat belt is thousands of miles inland within a huge continental area, whereas most of the examples of circles in other countries seem to occur in cereal crops nearer to the coast. In Canada there are thousands of large bodies of water in the Great Lakes system which might help explain the prevalence of circles in their records. Geo-graphical and climatalogical factors are important considerations.

Hoaxing

The sceptics' usual argument to account for the circles is that they must be the result of human engineering – a polite term for hoaxing. The admission in September 1991 by the two artists that they had created hundreds of the more complex designs during the 1980s right across southern England was predictably greeted with scorn by many cereologists. The artists do seem to have evidence for their claims and most sensible people had already accepted that these bizarre patterns simply had to result from trickery. There were smiling faces and messages in English, such as, 'We are not alone.' Sadly, the creative hoaxers wishing to express that particular message forgot that, had the originator really been an extraterrestrial, he ought to have written, '*You* are not alone', not 'we'.

Cautious researchers, such as those working with BUFORA, had exposed hoaxes as far back as July 1983 when a national newspaper paid a farmer to let them create a five-circle formation. They did so hoping to fool another newspaper into believing the circles were genuine, but the other paper had lost interest in the subject. The hoaxers themselves never revealed the truth until they were exposed by research put together by Ian Mrzyglod and Philip Taylor. Faking a circle is very easy. Some people simply use a central pole and a rope, others spread the load using planks or blocks of polystyrene foam, thus crushing the crop less badly. Even large circles can be swirled flat in minutes.

Once the publicity began and newspapers started to offer large cash prizes for the best circles, theories and stories, the incentives increased. However, as there were now thousands of circle watchers in the fields each summer, with cameras and binoculars ready to catch the 'intelligence' in the act, the risk of faking a circle increased. During 1990, 1991 and 1992 there were several cases of hoaxers being spotted and even filmed doing their deed. The most notorious of these occurred during the infamous 'Operation Blackbird' site watch organized by Colin Andrews and Pat Delgado with support from global television companies. They had provided image-enhancing cameras that recorded thermal radiation even in the dark. When the cameras captured lights one night

The Mowing-Devil:

Or, Strange *NEWS* out of

Hartford-ſhire.

Being a True Relation of a Farmer, who Bargaining
with a Poor *Mower*, about the Cutting down Three Half
Acres of *Oats*; upon the *Mower's* asking too much, the *Far-*
mer ſwore, *That the Devil ſhould Mow it, rather than He.*
And ſo it fell out, that that very Night, the Crop of *Oat*
ſhew'd as if it had been all of a Flame; but next Morning
appear'd ſo neatly Mow'd by the Devil, or ſome Infernal Spirit, that no Mortal Man was able to do the like.
Alſo, How the ſaid *Oats* ly now in the Field, and the Owner
has not Power to fetch them away.

Licenſed, *August* 22th, 1678.

Left A woodcut from the seventeenth century which tells the story of a 'Mowing Devil' at work in Hertfordshire. Was this how crop circles were interpreted during the Middle Ages?

Below Physicist and meteorologist Dr Terence Meaden who was the first scientist to study crop circles

and as dawn broke a pictogram was seen emerging from the mists in the field below, the news was flashed instantly around the world with claims that something truly extraordinary had happened. Sadly, time revealed that the circles were a very crude hoax (complete with ouija board in the middle!) and the images that were picked up were not objects in the sky, as some had speculated, but more probably the body heat of the human hoaxers.

By 1993 serious doubts had grown that it was possible to tell hoaxed circles from real circles, especially if the hoax had been carefully carried out. A 'hoax a circle' competition in July 1992 was promoted by a newspaper and a cereologist journal; they were hoping to demonstrate that circles faked under optimum conditions would be different from the real ones. Unfortunately, the fabricated circles were at times so good that they almost established the case for hoaxing by themselves. One prize winner, Jim Schnabel, created a good complex pictogram alone in the dark.

Circular Explanations

As soon as crop circles began to be reported, a cottage industry developed around attempts to find explanations for their cause. While none have had lasting credibility they have ranged from the serious to the absurd. Animals have always been thought of as a popular cause. It is true that some creatures such as foxes can create circular depressions in a field when they make lairs for their young. However, these are very small and do not resemble real crop circles.

Rather more difficult to believe is the idea that circles are the result of rampant hedgehogs in a frenzy of love-making, although the image has given a few tabloid newspaper editors a smile or two.

A common view is that the similarity in outline between the circles and ancient stone circle sites and burial mounds must be relevant. Perhaps the foundations of these under centuries of earth cause the soil to collapse and lay the crop down in outline patterns. Sadly there is no evidence to support this clever concept and it struggles in the face of the data from overseas.

Among the wilder theories is the suggestion that World War II bombs dumped by fleeing German aircraft are decaying and exploding deep below the ground causing the soil above to collapse. In crop circle cases, however, the soil does not collapse.

Even more delightful is the suggestion that high-flying aircraft are depositing the contents of their lavatories, which freeze as they fall tens of thousands of feet to the earth, create a circle by impact and melt before the discovery is made. Amusing as this may seem, it is true that blobs of ice have fallen from clear blue skies on several well-recorded occasions and there are those who link them with aircraft toilets. However, there are too many examples of crop circles preceding the invention of the aircraft (*see* page 24.)

The most popular idea is that the circles are a message from an alien intelligence. Attempts to translate what they say have failed, but have been diagnosed as Atlantean, ancient Sumerian or even Venusian.

The weather theory

Dr Terence Meaden, a physicist operating a meteorological research bureau, was the first scientist to study crop circles. This was during August 1980. His specialist topic is tornado and whirlwind damage so it is not surprising that he perceived these marks as originating from such damage. Initially he suggested in a series of articles between 1981 and 1983 that gently swirling summer whirlwinds forming in the lee slope of hills were creating the patterns. However, when the circles became ever more complex from the mid-1980s onward he developed the theory that a new type of wind vortex which could shear on impact and produce multiple patterns like a tornado was the more probable cause.

As the circles got even more fantastic, he chose to adapt his theory to try to make it fit, to the point of seeking to explain the pictograms. Most scientists, unfamiliar with the facts of course, believed that if some circles were hoaxes then all of them must be. Few could accept that vortices, however novel, could gouge out formations with straight lines and artificial-looking shapes.

Towards the zenith of his escalating theory Dr Meaden wrote a book in 1991 suggesting that the ancient Britons were familiar with circles and built monuments in their image (*see* page 4). However, Terence Meaden does have growing support from physicists and meteorologists. His work has appeared in house journals such as *Weather* and in June 1990 a major international symposium was staged at Oxford on the science behind crop circles. It attracted professors from three continents to discuss the idea of what was termed an ionized plasma vortex. The ionization effect was the latest update to the theory and said to result from the rotating air as it spun towards the ground. It was thought to be the cause of glowing lights (i.e. UFOs) sometimes seen at crop circles around the moment of formation.

Among the principal advocates for Dr Meaden's idea are Professor John Snow from Purdue in the USA and two Japanese experts in the field of plasmas, Professors Yoshi-Hiko Ohtsuki and Tokio Kikuchi. All visited English circles, went home and found cases in their own country and conducted laboratory experiments to try to vindicate the theory. These experiments were a significant breakthrough. A computer simulation recreated the optimum conditions for crop circle formation and produced circles 'on screen'. A scale model of one of the sites where circles frequented was set up in a wind tunnel to test airflow. Here artificial circles were reproduced on a small scale in the laboratory.

However, the most remarkable success was at a Tokyo research university where they produced a real plasma vortex using high energy beams. This was then fired onto a metal plate holding a thin coating of recording medium. The vortex was filmed and appeared as a corkscrewing column of twisting blue plasma which briefly touched the recording medium and left behind a small circle on the plate.

With this research, the scientists persuaded the Tokyo underground railway system to give them access to one of their tunnels. It was theorized that in such confined spaces the combination of rotating air currents and electrical fields should create circles. Film of the experiment, records, and samples brought back to the laboratory further prove that small circles were indeed found imbedded in the dust around the subway lines. All these artificial and subway circles were of the very simple type. None of them were pictograms.

These experiments have conclusively established the credibility of the plasma vortex concept and strongly imply that it could create circles in crop fields. Few people know about this research but the Japanese are devoting millions of yen in an effort to harness the vortex energy. In Britain, where the theory was born, most people still ridicule crop circles as the result of hoaxing.

Thankfully, in 1992 Dr Meaden did accept that all bar a few simple circles were made by human beings. Perhaps when the media stop reporting the mystery and the hoaxers give up, then just a few real circles will appear. If they all conform to Dr Meaden's ideas, theory may become fact.

A circular summary

■ The earliest known crop circle traced in historical records occurred in a field near Assen in Holland in the year 1590.

■ The area with the most crop circles is Wiltshire, England, with over 1000. Many are believed to be hoaxes.

■ The smallest crop circle known is under 1 foot (0.3 m) in diameter and the largest formation over 1000 feet across and filled an entire field.

■ Crop circles have formed in forty countries from Australia to Zimbabwe. This includes the-then communist states such as Russia.

■ There are forty cases of people who have reported seeing a circle form. None claim that spaceships were responsible and all describe invisible rotating forces descending from above and lasting only a brief moment.

■ In the seventeenth century scientist Robert Plot first suggested that circular patterns in fields might result from air blasts fired at the ground – a theory later independently adopted by physicist Terence Meaden.

■ Scientific research has established a link between circle numbers and the presence of lee slopes of hills, stable weather conditions and frontal systems generated by nearby seas or large areas of water.

■ Despite media claims to the contrary, no reliable evidence has been documented about cell structure changes to crops or unusual radiation detected within the fields. Cereal harvested from circles is perfectly safe to eat. Some results announced in late 1991 suggesting otherwise were largely retracted several months later when problems with the experimental protocol were found. Further tests were ordered in 1992 but no startling results have yet been released.

Other energies

While the weather theory has a good deal going for it, scientific possibility is not the same as certainty. Geoff Wain, an electrical engineer, has suggested a different and fascinating theory.

He proposes that when crop fields ripen, the action of wind blowing through the tall wheat produces excessive static electricity in the form of friction. In certain locations (for example, in chalky outcrops) and during dry weather spells, this charge would not leak out but could be stored as it is in a capacitor. Eventually, when full capacity is reached, it would be discharged in one sudden burst of energy. This would be passed into the atmosphere through the crop stems, which would move apart as a consequence like identical magnetic poles pushing each other aside. The balance of forces would sweep out a circular swirling pattern.

It seems feasible that charges might be stored in the ground for long periods and then be triggered into a rapid discharge by a rotating

Authorities

■ *Colin Andrews* and *Pat Delgado* wrote *Circular Evidence* (Bloomsbury, 1989), the international best-seller that first alerted much of the world to the subject. Delgado became interested in 1981 and started to write for UFO magazine *Flying Saucer Review.* He was joined by electrical engineer Andrews in 1985 and later they created a circles research team. They both supported a concept of 'unknown energies' triggered by 'unknown intelligences' but after a shorter sequel to their book in 1990 they went their separate ways. Andrews gave up his job to become a full-time circle researcher and in 1992 based himself in the USA to lecture and write on the topic. He seemingly became more closely associated with UFO ideas and in summer 1992 worked on a project in the English crop fields with a team of American researchers who had tested equipment at a UFO window area in Florida. This used lasers in an effort to trigger alien contact. In 1992 Delgado wrote *Crop Circles: Conclusive Evidence?* (Bloomsbury) updating his views, which remained much the same and showed little obvious concern after the Doug and Dave hoax claims.

■ *Paul Fuller*, a statistician from Hampshire, became BUFORA's circle consultant in 1985 and compiled a series of reports with Jenny Randles, the organization's Director of Investigation. Randles had begun work on circles in 1982. They aimed to disprove any link between UFOs and circles, but found that if Meaden's ideas were valid the energies involved might resolve some puzzling UFO cases. They compiled extensive records of pre-1980 circles and scoured the UFO archives for relevant cases. In 1986 they produced *Mystery of the Circles*, the first public report on the subject. This was expanded into a book, *Controversy of the Circles* in 1989 (BUFORA). In 1990 they wrote a book seeking to justify why the Meaden theory was significant to the UFO data and to show that the public had been misled by the hyping of the circle data. *Crop Circles: A Mystery Solved* (Hale) made only modest inroads into the success of the cereologists movement but was the first book to be translated into eastern Europe. In 1993 the book was revised and updated to discuss latest revelations. After first publication of their book, Fuller launched *The Crop Watcher*, a magazine to update findings, which reports extensively on historical and global cases as they are uncovered and seriously debates the hoaxing option.

■ *Dr Terence Meaden* was the first scientist to study circles, writing twelve articles in meteorological journals attempting to demonstrate his weather theory. In 1989 he published his own book *The Circle Effect and Its Mysteries* (Artetech). This did at least set the theory before a wider audience but had minimal impact alongside *Circular Evidence.* In 1991 with Oxford scientist Dr Derek Elsom he edited the proceedings of the conference on crop circles in June 1990. This book *Circles from the Sky* (Souvenir) contains a range of papers, including some scientific data on research at various universities but also has some more popularly accessible material, such as a lengthy paper linking UFO cases to crop-circle energies from the BUFORA research team. In 1991, 1992 and 1993 Meaden published a series of books with Souvenir beginning with *Goddess of the Stones* which examined the links between prehistoric man, his belief systems and the importance of the circle symbolism, suggesting that there may have been prehistoric awareness of the circles effect.

■ *Ralph Noyes* is a former diplomat at the Ministry of Defence who for a time ran the department which, among other things, collated UFO sightings. On retirement he became interested in strange phenomena and is well respected with close affiliations both to BUFORA and the Society for Psychical Research. He became one of the founders of the first nationwide crop circle societies – the CCCS (Centre for Crop Circle Studies), formed in 1990. Later that year he edited a book based on the original formative meetings of the organization entitled *The Crop Circle Enigma* (Gateway). This is the most wide-ranging of all titles yet produced, offering the opinions of the new age mystics alongside scientists such as Meaden. In terms of variety of approach and thought-provoking material the book has gained a good reputation among circle researchers.

- *CCCS* c/o Box 145, Guildford, Surrey GU2 5JY
- *The Cereologist* c/o 20 Paul Street, Frome, Somerset BA11 1DX
- *The Crop Watcher*, 3 Selborne Ct, Tavistock Close, Romsey, Hants SO51 7TY
- Earthquest can be contacted at Box 189, Leigh-on-Sea, Essex SS9 1NF

column of hot air – Meaden's fairweather whirlwind. These whirlwinds are commonplace in hot, stable weather conditions. They are visible in shopping centres sucking up loose debris on the ground. Just as the whirling air would touch a crop from above, it might set the electrostatic field in motion.

This is one of the more sensible explanations and might account for odd sensations sometimes alleged by visitors inside circles who claim that they hear buzzing noises and feel their skins tingling and hair standing on end. This is exactly what you would sense if exposed to a strong electrostatic field.

It is true that most circles do form in summer weather and in locations which are well suited to the generation of static charges. Further, these electrical charges might form the glowing lights seen near circles and interpreted by many as being UFOs.

However, many different types of terrain have generated what look to be identical circles, ranging from grass to rice paddy fields, snow and even wet tarmac on a roadway. It is difficult to see how electrostatic forces might be responsible for these. Circles have also formed during cold weather, mist, fog and even torrential rain. These might all be hoaxes, of course, but at present we have no way of knowing.

WINDOW AREAS

Through the window

The Earth is a strange planet that is constantly bombarded by radiation from outer space. It also seethes with its own energy fields created by the magnetic core and its electrical potential. In addition, there is evidence for more localized effects, which Dr Michael Persinger, a neurophysiologist from Laurentian University in Canada, has called transients. Serious research is being carried out into the possibility that these may from time to time interact with both the environment and human consciousness to stimulate all manner of paranormal happenings, from crop circles to UFOs.

This leads directly to the idea of a window area, something first introduced in the mid-1960s by John Keel, a New York journalist and maverick researcher. A window is a specific region where paranormal events cluster far more frequently than chance should dictate. Such regions can be traced back into history. There may be legends associated with witchcraft, dragons, monsters, demonic forces, and strange lights. These legends are clues that suggest the location is worth studying more thoroughly to seek out patterns of data. Although windows seem to be sporadically active, erupting rather like volcanoes or earthquakes and then laying dormant for periods, there is a long-term continuity of activity that singles them out from random events elsewhere. Indeed this geophysical analogy may be more important than it seems when it comes to interpreting just how window areas operate.

Another interesting point is that windows often have names associated with them which hint at their supernatural background. The American researcher, Loren Coleman has labelled them 'devil names' and they do often have this connection. However, they might also be more obscure, using local terminology, such as Manchester's Bogart Hole Clough – *bogart* being an old name for a ghost.

Into wonderland

In north Cheshire there is a typical window area surrounding a rocky outcrop called Helsby Hill. The precise focal point seems to be south of Warrington and Runcorn in an area near Daresbury and Preston Brook. On the surface this looks a quiet, peaceful spot with rolling fields near the River Weaver and the

Authorities

■ *Paul Devereux* has produced probably the most important book about window areas (along with fellow researchers David Clarke and Andy Roberts and geologist Dr Paul McCartney). *Earthlights Revelation* (Blandford, 1989) surveys the key window areas worldwide and has photographs, some in colour, taken on location. It also explains the latest thinking from earth mysteries researchers about the meaning of these areas.

■ *John Keel* originated his research into window areas through his study of the sightings of bird-man like monsters in Ohio subsequently published in *The Mothman Prophecies* (E. P. Dutton, 1975, reissued 1991). His witty and often tongue in cheek reflections have appeared in several further volumes. *Strange Creatures from Time and Space* (Sphere, 1976) is full of window area stories and his recent *Disneyland of the Gods* (Amok Press, 1988) is partly a compilation of his articles for *Saga* magazine.

■ *Larry Kusche* has produced the best book about the Bermuda Triangle. In it he demonstrates how many of the cases are open to more rational interpretation. The book is based on diligent research through the official archives of each missing ship and aircraft. After *Bermuda Triangle Mystery – Solved* (Harper and Row, 1975) little mystery remains.

Books about individual window areas are fairly rare. A few worth referring to are:

■ *Project Hessdalen* (c/o Boks 14, N-3133 Duken, Norway) has detailed reports, including colour photographs and maps, of a major Scandinavian window.

■ Randles, Jenny, *The Pennine UFO Mystery*, (Grafton, 1983) looks at the Todmorden/Bacup area of the Pennine Hills.

■ Hough, Peter and Randles, Jenny *Mysteries of the Mersey Valley* (Sigma Press, 1993) has more detail on the events in the Daresbury window area.

■ Stacy, Denis, *The Marfa lights*. This discusses a key American site. c/o MUFON, Box 12434, San Antonio, TX 78212, USA.

estuary of the River Mersey. But for centuries it has attracted strange phenomena in great abundance.

Long ago it was noted for its outbreaks of poltergeist activity which legend has it was caused by a dragon set loose in the area. In the twentieth century the area has had a bewildering range of events on its doorstep. There are no fewer than six well-attested cases of vehicle interference within three or four miles of Preston Brook. There are only about six hundred such events reported worldwide so this local activity would seem statistically very significant.

In a typical example a motorcyclist was driving north towards Daresbury late at night when he encountered a light overhead and stopped on a sharp bend. The next thing he knew he was some miles away, near Chester, without any idea how he had reached this isolated spot.

In another example, a car driving past the village of Daresbury was suddenly 35 miles (56 km) north near Preston in Lancashire. There was no apparent transition. The distance was travelled in a moment.

Preston Brook has also had sightings more recently of monsters and two UFO abduction claims, with weird noises heard in the night, strange lights in the sky, poltergeist outbreaks and one of the county's best crop circles.

There was also a remarkable case in January 1978 when four poachers beside the River Weaver saw little entities emerge from a bell-like object and attempt to place some immobilized cattle in a big cage. They were at a spot known locally as the Devil's Garden.

This place is quite a wonderland, which is very apt because Daresbury was the original home of Reverend Charles Dodgson (Lewis Carroll) whose timeless tales tell of Alice's adventures in such magical realms.

The Bermuda Triangle myth

The Bermuda Triangle is the most famous example of what many think of as a window area. This is, however, more legend than reality.

It is true that ships and aircraft disappear in considerable numbers while crossing a huge area of Atlantic Ocean between the coast of Florida and the Azores. But this is one of the densest transportation zones in the world, so it has to be seen in relative terms – more ships vanish because there are more ships to vanish.

Silbury Hill in Wiltshire is a good example of a window area. Strange phenomena such as mystery lights, crop circles and odd sound and electrical effects have frequently been reported here

The so-called triangle is not really a triangle. Different authorities ascribe to it a wide variety of geometric forms covering varying areas of sea. Some are so large it would be hardly surprising they had a number of puzzling disappearances within them.

Close scrutiny of some classic cases from the catalogue of triangle mysteries shows that very often crucial facts are omitted from popular versions of the tale: a ship may have been carrying explosive cargo or there may have been a sudden violent storm. However, there do seem to be interesting anomalies that have been reported in a small region east of Bermuda. These appear to be connected with electromagnetic fluctuations. This spot may be a true window. Several ships and aircraft have reported losing compass headings and radio contact when they pass by. Aviation expert Martin Caidin told the authors that he discovered that a NASA shuttle had experienced similar problems when traversing this region of air space.

Caidin himself, during a flight from Florida to Europe lost all communications on his aircraft despite its new multi-million-dollar satellite navigation link which had been installed recently. For more than an hour he and his crew flew blind in a strange cream-coloured sky without sign of any horizon. Then, suddenly, they burst out into blue sky and resumed communications.

What are the windows?

If these window areas do exist then what brings them into being? A popular theory (indeed the very term 'window' implies it) is that they are like portals to another reality, possibly a different dimension of space and time. Phenomena from that different reality may step through and we may slip out of our own world for a brief moment. This could make sense of some experiences, notably those where a witness describes how they seemingly entered an alien environment parallel with, and closely similar to, our own, but clearly

Window reflections

- There are many isolated locations around the world where strange phenomena focus far more strongly than chance should dictate. Most countries have several of them, each being perhaps a few kilometres square. These were named 'window areas' by American writer John Keel and the term is now widely applied.

- Research has suggested that these are locations where natural energy forces are concentrated, bringing with them an assortment of seemingly paranormal phenomena.

- A key factor appears to be the presence of certain types of rock, especially containing crystalline structures such as quartz which can store and emit electrical energies if placed under strain.

- Trigger factors isolated by researchers such as Dr Michael Persinger in Canada and Paul Devereux in Britain appear to be reservoirs, dams and other construction work, rock excavation, and even the passage of heavy air pressure systems which can 'squeeze' the rocks.

- The generated energy can 'glow' to produce light phenomena, usually interpreted as UFOs today but given the generic term 'earthlights' by researchers. There may be local legends associating these lights with the window area across many centuries, alongside other seemingly paranormal events. History may even recognize that fact, for example, by including references to 'devils', 'demons' or 'leys' in place names.

- The lights have been successfully reproduced in laboratories in the UK and USA by recreating mini window areas.

- There seems to be a link with seismic activity. Earthlights may help defuse tremors by 'leaking strain' gradually. A possible route towards limiting major earthquake activity may emerge from this work.

not quite normal, everyday reality.

Researchers call this fugued state the Oz factor, after the classic fairy tale *The Wizard of Oz*, where a young girl is suddenly transported from her Kansas home into a strange reality and finds here that magical things can happen. It is not uncommon in a window area for witnesses to report that the people or traffic that should have been surrounding them were completely absent for the duration of the experience. These returned only after the experience was ended. This is rather like stepping into a parallel universe – one which almost, but not quite exactly, duplicates the one the witness has left.

However, there are other possible ways to account for the Oz factor. Witnesses to a range of paranormal phenomena often say that before strange events begin their thoughts 'go inward on themselves' and they stop paying attention to the outside world. They may even cease talking to one another. It is not unlike a self-induced sensory deprivation.

In an incident in August 1992 a family was driving to a shopping centre in Buckinghamshire on a Saturday morning, when they suddenly encountered a strange bank of mist. They found that not only had they stopped singing Beatles tunes but that nobody was talking to one another. This was most unusual given that two young children were in the car. All the traffic seemed to have melted away and ambient sounds had faded. Then they found themselves further down the road from Hockliffe without awareness of how they had reached this point. When they got out of their car all their senses were 'out of phase' and they could not grab the door handle properly as if they had temporarily lost coordination.

While we do not know what really happened to this family all the symptoms they describe are very typical of the Oz factor.

If Persinger is correct and floating masses of electrical energy do exist, then this bank of mist which the Hockliffe family encountered could have been a transient. It may be that the electrified mass is attracted to the metal body of a car. Persinger's work is amplified by earth mystery researcher Paul Devereux and they believe these transients (or earthlights as Devereux calls their glowing nocturnal counterparts) would congregate in geologically active areas. Certain types of rock below the surface, fault lines causing pressure or reservoirs and quarries on the surface may destabilize the delicate balance of forces within the crystalline rock. As a result gases or electrical energies would be released which might be called UFOs. Some amorphous shapes may look like entities or monsters. Indeed Devereux even suggests that shapes may change in response to possible psychokinetic abilities of human consciousness, effectively moulding the transient into a 'living nightmare'.

Of course, the existence of the released energy would also be related to some outside influences, such as fault-line movements, changes in the Earth's own magnetic field, passing weather systems, perhaps even solar radiation storms – all have been found to provide some evidence of a link. In this way windows would be more active at some times than at others, producing bursts of activity, called 'flaps' or 'waves' after which the area might lie dormant.

Persinger's experiments into the way in which these electrical fields can trigger an altered state of consciousness offer a solution to the strangest of paranormal phenomena within these regions as well as to the Oz factor itself. A witness may simply enter a somnambulistic state where they can temporarily disassociate from waking consciousness and either dream, hallucinate, lose consciousness or even drive under a sort of 'automatic pilot' just as a sleepwalker would act when not consciously awake. The latter is a well-known psychological phenomenon called 'highway hypnosis' leading car drivers to lose miles of a journey from their recall. The result of these processes would be visionary experiences, sudden apparent missing periods of time or even spatial transfers from one location to another. These are exactly the most common events reported within window areas.

SPACE INVADERS

FALLS FROM THE SKY

Small creatures, such as fish, have mysteriously tumbled out of the sky throughout history. This illustration is taken from Lycosthenes' *Book of Prodigies* which appeared in 1557

What goes up . . .

There is an old saying that whatever goes up eventually comes down. Sir Isaac Newton (1642–1727) discovered this fact, as legend has it, when an apple fell onto his head. This even led him to understand the force of gravity. However, something rather stranger than mere gravity is involved in the mysterious phenomenon of unexpected objects falling from the sky. These are sometimes called 'fafrotskies', a name derived from the words 'falls from the sky'.

Many ordinary things do, of course, fall out of the sky in perfectly understandable circumstances. Rain is formed when air rises carrying water vapour which turns into droplets forming a cloud. As the droplets merge together they form large drops too heavy to float and they cascade to the ground. However, there are some things that drop from the sky for which science has no immediate explanation to offer.

John Lewis, who was sawing some wood at Mountain Ash in south Wales on the wet morning of 9 February 1859, was astonished when little objects began to flop onto the ground around him. They were tiny fish just a couple of inches long. There were not simply one or two of these unexpected arrivals, but hundreds of them. They covered a strip of land two hundred feet (61 metres) wide and although he and a friend rescued some by placing them into a rainwater pond, most others died. The largest he found was 5 inches (127 mm) long. A local specialist had a look at some survivors and found nothing unusual about them. They were fish indigenous to nearby rivers. But how had they fallen from the sky in this dramatic way?

Fish, sand, ice blocks and unexpected blobs of goo are relatively common fafrotskies. But there have been coins, beans, peas, tiny frogs, and in one case what might almost amount to manna from heaven.

On 17 March 1992 Doug and Paula Wood from Bellingham in Washington, USA, had the shock of their lives when 20 lb (9 kg) of white dough similar to that used to bake bread literally crashed down through the roof of their home with a noise described as akin to a sonic boom. Local Christians immediately picked up the connection with the Biblical tale of God sending sprinklings of manna from the sky to feed the chosen Israelites. But the Woods' free gift was diagnosed as pretty ordinary stuff, which rapidly turned into an inglorious slimy mess. Only two mysteries remained about it. Where on earth did it come from and exactly how did it smash its way to earth like a gooey meteorite?

Within these two questions lies a puzzle that has baffled researchers for hundreds of years.

Drop-ins from the past

You can find strange falls reported in many historical records. Greek scholar Athanasius told how small fish were seen to come down for several successive days in the Chersonesus region during the fourth century AD. Another scholar, Athenaeus, reported regular falls in certain locations throughout the previous hundred years. It was quite a commonly remarked-upon phenomenon even by the time of the arrival of Christianity.

Indeed there are those who argue strongly that the manna from heaven reference within the Bible is more than an allegorical tale but is really a genuine example of this scientific mystery. Indeed it is known that some types of moss and lichen can be carried aloft by high winds and drift for days in thermal layers. When released these are capable of producing a residue that, if mixed with water, can provide a tasteless but quite edible foodstuff. Perhaps this was the cause of a persistent and eventually romanticized story which has become the famous account that we are familiar with today.

From time to time in the past couple of centuries newspapers have picked up stories of an assortment of unexpected fafrotskies. More recently they can be so well attested that there is little doubt as to what had happened. When a shower of live frogs fell on a town in Wiltshire in 1939 it was reported by the London *Times* that the streets were so choked that it was almost impossible to walk without standing on them.

The phenomenon occurs all over the world. The residents around Quito in Ecuador, South America, believe that the local volcano is responsible for their fish falls. Indeed legend has it that on the night of 19 June 1698 when a whole mountain top disintegrated in a gigantic blast, miles of fields were deluged with small fish.

In Honduras there is even a tradition built around the mystery. The villagers of Yoro take to the streets with buckets on the first day of the rainy season in the hope that they will be able to catch some live sardine. On a number of occasions this most unusual fishing expedition has allegedly proven possible. Nobody knows why.

Of course, different cultures have developed their own mythologies to explain the mystery. Not uncommon is some form of sky god that rewards a tribe for its peacefulness during the winter and as a consequence showers them with food prior to the start of spring. Other concepts have ranged from underground devils demonstrating their dominion over life and death, to the showers sent by ancestral spirits in some regions of Africa, who wish to bestow blessings on their surviving relatives and ensure the continuation of the harvest.

In most modern Western cultures we have replaced such legends with the mythology of science. As such everything from tornadoes to earthquakes has been cited as the trigger for some sort of upheaval that casts poor creatures into the sky and eventually sends them crashing down again. The problem is that few of these explanations really try to describe how the process might work. They merely look for a way out of an inexplicable circumstance. They also offer little indication of how the fish or frogs could possibly survive their journey into the blue and then the long and traumatic trip back down again, often with no ill effects at all.

Some latter-day fishy stories

Sardines it seems fall from the sky with alarming regularity. At noon on 6 February 1989 the town of Rosewood in Australia, on Queensland's Gold Coast was at the heart of one such mystery. Harold Degen and his young son were returning home for lunch amid a hail-like rainstorm. But all of a sudden this turned from ice to dead sardine, dozens of them 'like silver rain', as they were described. When Harold told his wife what was happening she did not, of course, believe him until looking for herself and finding them bouncing off the ground. Then a desperate search for a solution was made. Could they have been thrown into the air by passing traffic? But as more fell, creating a thick carpet across many square yards, this was obviously impossible. Overflying birds quickly spotted the windfall and had a feast. Indeed, some commentators suggested later that marine birds were the source, dropping the fish from their bills as they passed overhead. However, that seems unlikely as so many specimens were involved over such a localized area. Meanwhile, the mystified Degens kept a few of the aquatic wonders to feed their cat.

While sardine descending onto a near coastal town is baffling, but not utterly inconceivable, a similar occurrence miles inland is the source for rather more consternation. On 22 September 1991 at Plaistow, Greater London, five sardines were found lined up in a precise formation across the concrete path in a woman's garden. She did not see them arrive, but they were dead yet hardly damaged, just as were those on the other side of the world in Queensland.

The fish in Plaistow were about 4 inches (102 mm) long and the Natural History Museum in London confirmed to researchers with *Fortean Times* magazine that they were sea fish not found within the River Thames. So how had they reached east London? It was suggested that they were from a fishmonger's and had been cast over the wall, perhaps as a joke. But the woman who found them disputed this idea. Researcher Bob Rickard also noted how their formation oddly resembled fish swimming in a shoal.

The return of Mr Blobby

On the afternoon of 16 June 1979 in Mississauga, Ontario, Canada the Matchett family were in for a real shock. Their teenage daughter, Donna, was raking leaves from the swimming pool on this hot summer day when a

Small fish that fell out of the sky on East Ham, London, in May 1984. Where did they come from and how did they survive the descent relatively undamaged?

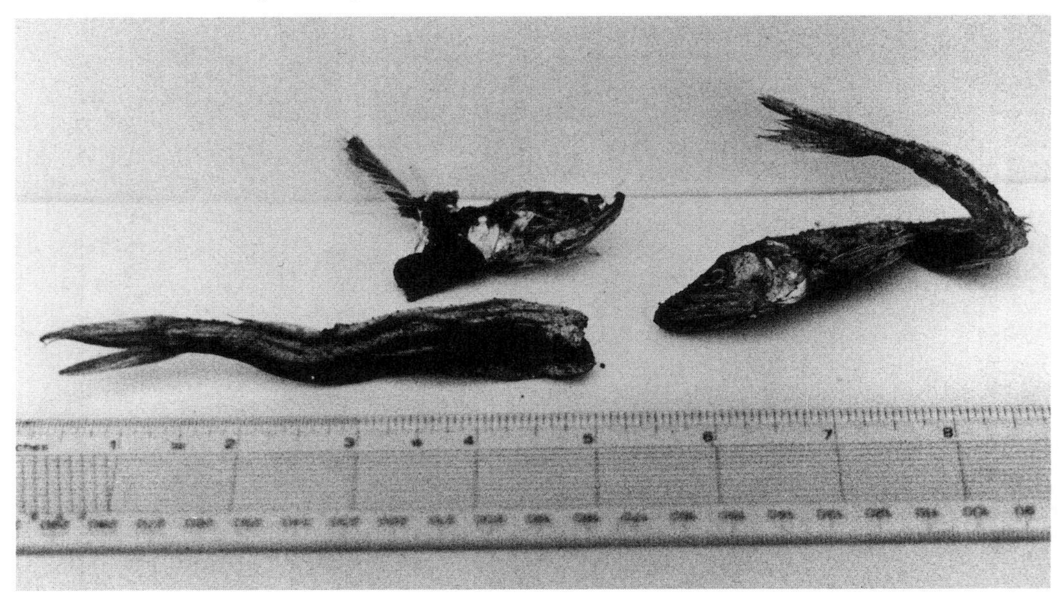

crash alerted her to something amiss behind her back. Turning around she spotted a bizarre green mess that had fallen onto a picnic table only feet away from her. Rising from this blob was a column of fire, like a jet from a blowlamp. Donna reacted quickly and used a hose to extinguish the flames leaving only what came to be known as 'the Mississauga blob' to mystify even seasoned students of the fafrotsky phenomenon.

As soon as the fire was put out the gooey mass solidified into a hard green flat-surfaced object riddled with pock marks and weighing a few ounces. The Matchetts tried very hard to figure out what this blob might have been. The Ministry of the Environment took part of the object for analysis and concluded it was a common type of plastic used to make objects such as toys. It had been heated, melted and resolidified. As a result of this the Ministry came up with one of the least likely fafrotsky explanations so far – that an incendiary device or flare had struck a child's frisbee in mid-flight, set it on fire and deflected it like a missile onto the Matchett's garden.

Meanwhile the police also carried out a test at the Forensic Science laboratory in Toronto. They found broadly similar conclusions to the government research team, adding that due to the combustion no precise identification could be made. However, they said that the report was confidential and, according to investigator Dwight Whalen, allowed the Matchetts only to read the analysis but not to take a copy.

Traven Matchett, dissatisfied with the frisbee theory tested it himself by trying to set one alight with a blow-torch. It took several minutes to do so and then only smouldered. Clearly, he argued, this was not the cause.

Another theory was that the morass had come from outer space, possibly residue from a crashing satellite. This idea seemed inconclusive when it was revealed that three other blobs of plastic material had fallen onto gardens in surrounding towns during previous months. None was quite so spectacular, bar one at Brampton, Ontario, which was three times larger than the Mississauga object.

Falls such as these are fairly rare but do continue to happen. On 26 August 1992, in the wake of a hurricane, some horrible gooey masses were found on the land of Elwood Guillot near New Orleans, Louisiana, USA. The blobs resembled oily cowpats, although slightly larger in size. A biologist who analysed them said that were single cell organisms but nobody could determine their origin. Upon searching a nearby pond several more were discovered and it was also found that they tended to evaporate slightly when left exposed to the sun.

In late 1992 UFO investigators Ken Phillips and Judith Jaafar were handed an object that seemingly had plummeted out of the sky into the garden of a couple in the Quantock Hills of Somerset, who then became embroiled in a complex close-encounter case. Jenny Randles had the opportunity of examining this: it was a small jade green rock, a few inches in diameter, with a smooth glassy surface and pock marked holes. The couple naturally feared that it might be contaminated but put it in their fish tank for safe keeping. Upon inspection by the Geological Museum in London it turned out to be ordinary plastic ironically used for ornaments placed in fish tanks.

It is wise not to make assumptions about the nature of these falls. A slimy gooey mess was found in the middle of a swirled crop circle in Wiltshire in July 1986 and was subject to speculation that it might be connected with whatever force forged these strange patterns in the ground. But analysis revealed it was a mixture of plant cells and animal fats and was probably a chocolate bar dropped by someone inspecting the circle. The melted mess had mixed with crop and decomposed into the resultant slime.

Oddities from space

Objects fall from the sky all the time. Jenny Randles was within seconds of a first-hand example in early 1992 when an object crashed into the garden of a neighbouring house at Edgeley, Cheshire. It was several inches long and was identified later as a bolt used to lock aircraft wheels tight. These are removed before take-off but this one had flown across the Atlantic to drop from about 1000 feet (305 metres) during the jet's final descent into Manchester Airport. Jenny Randles' house was directly on the flight path a few hundred yards from the spot where the bolt embedded itself with ferocious impact.

One of the most disturbing instances of fafrotskies is the plague of tiny frogs that can accompany some rain storms. On 27 July 1979 at Bedford a small area was found seething with living tiny examples, no larger than a coin, hopping about as they fell to earth. In May 1981 a similar rain struck Nauplion in Greece and meteorologists in Athens were said to have identified the frogs as having come from marshlands in Africa thousands of miles away. How they had made this aerial journey unscathed was less of an immediate concern to the village locals than the fact that they were being kept awake by the croaking of this swarm of invaders.

On 12 February 1979 a man was sitting in his conservatory in Southampton when he was astonished to find it raining plant seeds. A wide range of seeds fell during the next two days, including mustard and cress, peas and beans. All of them sprouted natural plants after they were allowed to germinate.

At Thirsk in North Yorkshire on 18 June 1984 a torrential rain storm brought with it an unusual cargo of hundreds of tiny winkle shells. An inspection of them showed that the animals inside were still alive and that their closest possible origin was the North Sea some 30 miles (48 km) away.

On 5 June 1983 countless pieces of coke, some over 2 inches (51 mm) in diameter, fell onto Bournemouth and its environs.

Swansea in Wales went one better on 23 September 1984 when small glass beads fell onto the ground. At first a geologist from the university suggested that they might be debris from a volcanic eruption that had circled the world high above the ground, but when samples were analysed they were found to be man-made metallic fragments that had been melted. This did not explain where they had come from, of course.

Possibly the strangest mystery struck Hamilton, New Zealand in 1974. Cans and bottles descended onto a group of houses for four hours and despite a close watch by police officers with patrol dogs no mundane source for the bombardment was discovered before the bombardment ceased just as suddenly as it had begun. Local Maori leaders were not so bemused. They advised that spirits of the dead had been offended by local residents and were getting their own back by dumping this celestial garbage onto their heads.

Sands from the sky

During the mid 1960s the phenomenon of 'dark days' around Lancashire, England, created quite a scare. The same was true in other parts of Europe. For a short time the sun would be almost completely blotted out and mid-afternoon on a summer's day would be turned into the blackest night. There were fears about the end of the world, but a clue about the true explanation emerged when it was discovered that cars in the black-out zones were found to have been covered in a sprinkling of a yellow dust.

The solution that became accepted by science was that sand from the Sahara desert in Africa had been whipped up by fierce wind storms and trapped by stable layers of air high in the atmosphere. This had then traversed thousands of miles and eventually been deposited back to earth by the different climatic conditions above northern Europe. Rains of sand have since become a more regular feature. In the summer of 1984 Europe was literally blitzed by them. Orange rain fell around southern Spain on 19 June. By August the clouds

Not only fish fall out of the sky. Showers of small crabs – usually alive – are common. This spider crab descended on Brighton in June 1983

FATE magazine has been a frequent reference source for mysterious falls. A 1958 cover depicts a terrifying shower of live frogs which fell on the USA

reached Dorset in England and were seen to resemble tiny grains that crumbled into powder when touched. The Sahara storm theory was proposed, but on this occasion a spokesman for the meteorological office in Bournemouth had to admit that the wind directions at the time were not consistent.

In 1987 a shower of tiny pink frogs (all alive) fell on 23 October in the Stroud area in Gloucestershire. A nature conservation group in Gloucestershire investigated two other outbreaks of frog falls which were alongside sand rains and the idea was proposed that the mini animals were transported by the winds in drops of water along with the clouds of sand. The image of thousands of frogs making such a day-long aerial hitch-hike miles high in the atmosphere and yet without suffering any ill effects does sound incredible. None the less, the identification of the frogs as an African desert species was announced after an enquiry. Their pink colour is a camouflage effect to blend in with the crystalline sands in which they live.

Whether the Sahara sand is a real clue remains uncertain. But the falls continue to happen. In March 1991 a frontal system over southern Germany and later Helsinki in Finland both produced the typical yellow detritus which covered cars and houses. Interestingly this fell as yellow snow in the more northerly regions. The drift was over a foot thick in parts of Lapland and analysis showed that it was sand consistent with the type found in Africa that created this unusual pigmentation.

Similar clouds of dust and sand sucked high into the atmosphere and circling the earth are

In May 1959 a chain fell out of the sky on top of a tractor at Rock Hill, Missouri, adding one more weird story to an ever-growing list. The investigating police officer looks suitably perplexed

thought by some scientists to be a solution to a number of puzzling supernatural events. In the alleged miracle of Fatima in Portugal, in October 1917 the sun dulled and rotated in the sky before countless witnesses. This may have been the result of a temporary dust blockage. The expression 'once in a blue moon' may owe its origins to real occurrences resulting from the effect of such sandstorms in the sky.

Ice bombs

Undoubtedly the most dangerous fafrotsky is the ice bomb. This is a remarkably common phenomenon and its destructive power has been demonstrated more than once by huge holes smashed through the roofs of houses. Yet, remarkably, despite many well-attested incidents, including one where a sheep's neck was allegedly severed by such a fall, there is no known case of anybody being hit by an ice bomb and suffering serious injuries. But there have been some very narrow escapes.

One example is that of the close encounter of Thomas Andrews from Easton, Maryland, USA. On the afternoon of 8 September 1984 he was in a caravan park sitting out in the open talking to friends when there was a whooshing noise. Masses of green coloured ice were being scattered all over the ground by an impact from above. The main mass was about the size of a basketball. Fragments were collected but melted into a vile smelling green liquid.

Colorations such as this are not uncommon. Blue is a frequent hue for ice bombs suggesting to many that some kind of chemical is the cause. By far the most acceptable explanation is that ice bombs are the content of aircraft lavatory systems. Although these are not designed to drop from the sky the frequency with which they allegedly do is alarming. However, to date no specific aircraft has ever been identified as a culprit, and in truth, there are few cases where this seems certain to be the answer.

A fall at St Julien-les-Villas in the Aube region of France on 8 July 1984 may be one of the rare exceptions. Here a five-foot (1.5 metre) hole was smashed through the roof of a house by a block of ice that fell from a cloudless sky. A chemical analysis showed that it was made up of 20 per cent urine. The house was also underneath a corridor where high-flying aircraft regularly pass.

In January 1990 Federal Aviation Authority experts confirmed that a Boeing 727 lost an engine in Florida in marshland when blue ice from its defective toilet froze, broke off and sheered it from its wing.

However, there are doubts about the universal acceptance of the aircraft toilet theory. The most serious is that in the 1920s, journalist Charles Fort collected dozens of ice-bomb stories from a time when high-flying aircraft did not exist and some stories even pre-dated the invention of the aircraft.

A case from Kazan on the banks of the River Volga in Russia suggests another theory. A man strolling on the beach at this holiday resort heard a whooshing, whizzing noise and was only a few feet from an ice block that imbedded itself into the sand in front of him. He took it back home and preserved it in his freezer for inspection by scientists from Moscow who concluded that it was a water-ice meteorite.

In May 1990 a glider pilot had just landed on Butser Hill in Hampshire when a huge ice bomb was spotted 500 feet (152 metres) up on a downward path. He was convinced it was a falling meteorite from space. It landed some yards away and exploded into fragments. He took the largest fragment away and kept it. This was remarkably smooth and about as large as a grapefruit.

The most intriguing case, however, occurred when Dr Richard Griffiths, a scientist, was almost hit by an ice bomb that ploughed into a suburban street at Didsbury, Manchester on 2 April 1974. This ice bomb was the subject of a detailed study from which very valuable information was gleaned.

There were fifty-one separate layers of ice in this large conglomeration. Air bubbles were trapped between them. This showed conclusively that the ice had formed from atmospheric clouds. This ice bomb seems to have grown like a snow ball or hail stone, layer accumulating upon layer. A single flash of lightning had been seen by Dr Griffiths just before the fall and, while no direct correlation can be established, some researchers suggest that the resultant electrostatic field might cause ice particles to be attracted together into a kind of super hailstone measuring up to several feet in diameter. The problem is how such an object could defy gravity and remain aloft long enough to form in the first place.

Authorities

■ *Dr Louis Frank* is a renowned American physicist who has created a storm by offering strong evidence to support his view that the Earth is constantly bombarded by micro-comets. These, he says, are made of water and ice and strike at the rate of up to many thousands per year. He argues that these are responsible for the unusual amount of water found on our planet and, thus, ultimately, for life itself. His thesis has generated a major scientific debate but is backed by photographic evidence from satellites of what he claims must be such comets. He also suggests that visual sightings of these comets when occasionally large enough to plunge through the atmosphere trigger some UFO reports and that cases of ice bombs may also be hard evidence of the concept. He cites Hu Zhong-Wei, an astronomer from Nanjing University in China, who has reported an analysis of three ice-bomb cases from the vicinity of Wu-Xi in 1982, 1983 and 1984. The final fall was available for detailed chemical and structural study and the results suggested that its origin may well have been from a small comet from outside the Earth itself. Frank has detailed his research in an entertaining and invaluable book *The Big Splash* (Avon, USA, 1991) which illustrates how anomalies that do not fit conventional science struggle to be accommodated by its traditionalists.

■ *Bob Rickard* is an Englishman who has for many years studiously collated stories of falls from the sky for his journal *Fortean Times*. Three compilations of early issues have been published covering the early, mid- and late 1970s and more are promised. These all offer a wealth of anecdotal accounts of numerous stories and are published by *Fortean Times* at 20 Paul Street, Frome, Somerset BA11 1DX.

Fall facts

■ All manner of strange things fall regularly from the sky – including foodstuffs, crops, material objects and even animals. These falls have been recognized for thousands of years and tend to be ascribed to a supernatural force. Possibly the most peculiar fafrotsky ever reported were three umbrellas that came from a blue sky in Durban, South Africa in 1980.

■ One of the most common types of falling object appears to be the 'ice bomb' – large chunks of ice descending to Earth from an often clear sky. They are regularly attributed to faulty aircraft toilet deposits freezing as they fall to earth, but reports long pre-date the birth of flight. The most recent scientific theory is of thousands of ice comets bombarding the atmosphere from the solar system on a permanent basis. The largest known ice bomb was over 5 feet (1.5 metres) wide and weighed as much as a man. It demolished the roof of a house in Devon, England, in June 1984.

■ For other types of falling objects the most popular theory is that whirlwinds suck them upwards and carry them aloft for long distances before depositing their content back to Earth. This struggles to explain the most bizarre cases where animals, such as fish, fall to Earth (often some distance from any water) and yet are still alive.

■ A popular theory proposed by some paranormal researchers, but scoffed at by science, is that these falls occur by way of a natural mechanism in the Earth's atmosphere that can by-pass space and time and teleport objects from one location to another.

From whirlwinds to comets

Ideas about the origins of fafrotskies are many and varied. Indeed there is every chance that different types of falling objects will require varying solutions. For example, it may well be that frozen debris from aircraft toilets is sometimes the cause of ice bombs. On the other hand sheets may be dropping straight off the aircraft wings when they are suffering de-icing difficulties. Bitter cold conditions are to be found at the height some aircraft travel through the atmosphere. In addition, as we have seen in the Manchester ice-bomb case, the more likely solution is that the object formed naturally rather like a hailstone.

However, there is yet another new theory which was announced in January 1992. Investigators from the air accident investigation team were puzzled by the fact that three aircraft on one day in August 1991 suffered sudden increases in wing icing which caused them to lose control. They were all in the same area of the Midlands in Britain at the time. The pilots reported a rainbow effect appearing suddenly in the sky. It is now believed that this was the result of sunlight shining through a huge field of otherwise invisible ice crystals that was somehow suspended in mid-air. Presumably, if such an icefield is shown to be a common occurrence within the lower atmosphere this may well turn out to be the origin of the ice bombs.

The ice may, of course, come from even higher, outside the atmosphere altogether and, as the Moscow scientists suggested, be from a meteorite or from fragments of comets that have sheered off. Comets are effectively huge 'dirty snowballs' that drift around the solar system and 'burn off' when they approach the sun, creating the trailing streamer or 'tail' for which they are best known. It may be feasible that some of this ice can make the passage through the frictional heat of the atmosphere and occasionally reach ground level.

However, all these possibilities concern the ice bomb. Falls of dust, seeds, and particularly, living creatures, need very different interpretations. The most widely speculated scientific solution is that everything from Sahara sand to sardines is picked up by whirlwinds and sucked into clouds, held aloft by air masses and eventually dropped back down to earth again.

There is a degree of precedent for this. On hot summer days fairweather whirlwinds are commonly seen. They are, in fact, a simple form of vortex akin to a tornado, but much less violent. You can see them in enclosed spaces, such as open shopping malls, where the rotating column of air is made visible by dirt and waste paper swirling round at speed and pulled upward. Farmers often have to confront them taking up loose grass, hay or corn from their fields and scattering it down again some distance away when the upward spiralling energy is counterbalanced by gravity.

In one case at Marple Bridge, Cheshire, on

15 June 1988 several witnesses saw hay being carried upward by such a whirlwind on two separate occasions a couple of hours apart. On the second, so much hay was taken into the air (so forming a crop circle on the ground) that it formed a solid lens shape at rooftop height that looked not unlike a flying saucer. After hovering in this position for a moment or two it began a horizontal flight across a school playing field and passed a church before disappearing from view. Moments later the energy holding the hay in the air seems to have dissipated and was seen by residents of a hillside housing estate. It was now spread out into a thinner cloud with several globules 'like galaxies in space'. These began to rain down on the ground, scattering hay for almost ten minutes over the houses, streets, gardens and a golf course.

While this phenomenon is interesting and shows how atmospheric forces can produce unusual rains, it is difficult to understand how these same whirlwinds could lift up much heavier objects such as frogs or fish and carry them for some considerable time.

The supernatural view

Those unsatisfied by the efforts of science to explain away strange falls from the sky have little to offer as an alternative. Some put forward an unproven and highly speculative science-fiction concept, that of teleportation.

In this theory the animals are unharmed by the passage of time and space because this occurs instantaneously. The fish or frogs, for example, are literally taken from their normal habitat and moved *en masse* to their new abode by passing through some sort of warp in time and space. It is true there are some intriguingly consistent stories where humans have been driving in cars and have suddenly confronted unexpected rain showers or low-level clouds. They enter these and then inexplicably report that they find themselves some distance from their original location without any recollection of how this has occurred or, in at least some cases, without any passage of time or use of fuel that would be consistent with them having travelled such distances.

One case is reported on page 34. However, the most remarkable such case involved a doctor and his wife driving south of Bahia Blanca in Argentina. They ran into descending cloud that straddled the road. They were with some friends in another car, who suddenly found that the doctor, his wife, and their vehicle had vanished. The couple say they 'woke' to find themselves on a strange road that they could not identify. This was hardly surprising as they were soon to discover they had been transported half a continent and several thousand miles to Mexico. An investigation of the circumstances made it seem virtually impossible that they could have driven this distance in the time.

If such reports do suggest that some form of teleportation is possible this might account for many of the features in the fafrotsky cases. It might also fit well with the way in which the sardines in east London were discovered – laid out as if in swimming formation.

However, there would still be difficulties. Why would the objects fall from the sky at all, rather than just be deposited on the ground? In some cases these creatures are never seen to fall but are simply found on the ground and assumed to have fallen there. Indeed, 'how else would they have arrived', witnesses usually ask? Why would the types of animals and fish be so frequently the same?

Perhaps time will tell, but the mystery of strange things that fall out of the sky looks set to continue.

BALL LIGHTNING

Bolts from the blue

Paranormal researchers often talk about ball lightning because it is a perfect example of how science struggles to accept unusual phenomena. For centuries people have reported experiences such as the following which occurred in 1920 at Parkside in South Australia.

A witness heard a hissing noise outside during a rainstorm. They went to investigate and saw a ball about 1 foot (305 mm) in diameter slowly falling from the sky and then down a fence. It was amorphous and vapoury with little shapes inside that looked like worms. It then hit the floor, shooting sparks off at the edges and

Ball lightning is not a modern phenomenon. Here Camille Flammarion records an encounter in France in 1845

proceeded to roll along a path, bouncing a foot or two into the air and squashing into an egg shape when it hit the ground. As it bounced upward for the final time, resuming its appearance as a sphere, it flew at speed through the sky and hit the wall of a nearby house, exploding into a shower of sparks with a loud bang.

One can well imagine how much terror such an incident can instil into a witness. Phenomena like this have been given many names – a common and very descriptive one being a thunderbolt. While such cases tend to be associated with thunderstorms and thus with ordinary lightning, surveys show that at least 10 per cent do not occur in storm conditions. Indeed some recent Japanese research by Professor Yoshi-Hiko Ohtsuki of Waseda University suggests that the percentage of non-thunderstorm ball lightning may be much higher than expected. One possible reason for this is that today many ball-lightning events are, understandably, misreported by their witnesses as UFOs and never find their way into the scientific database.

Fortunately, some physicists recognize this fact. Dr Paul Davies of Adelaide University in Australia has written papers noting how UFO records may be a treasure trove for science. Mark Stenhoff, head of physics at a London college and a leading figure in ball-lightning research, was an active ufologist for a time and is part of the meteorological team that has researched crop circles.

Nevertheless, despite the impact of individual reports such as the one outlined above, it took a long time for science to take to the subject. In 1950 Schonland stated in his classic review *Flight of the Thunderbolts* that many meteorologists and physicists felt that ball lightning did not exist. It was an optical illusion created by a brilliant flash of ordinary lightning imprinting on the retina of the witness and causing floating spots before the eyes.

Even today there is widespread consensus that this interpretation may often apply. However, it is somewhat difficult to relate such a solution to the Parkside case or others like it, particularly those where the phenomenon appears inside a room. Ball lightning is prone to this.

Spots appearing before the eyes is a totally inadequate explanation for the case of two RAF officers and their wives who experienced an encounter inside a bungalow at Katong, Singapore. There was a terrific monsoon which suddenly abated as if the eye of a hurricane were passing overhead. At that exact moment an object the size of a cricket ball and bright orange in colour appeared on the telephone wires outside. It looked fuzzy and made a hissing noise. It rolled along the wires, then jumped off to fly straight through the open louvred windows into the house. All four people then stared as the ball moved at walking pace 3 feet (1 metre) off the ground and passed across the kitchen. From a short distance they could clearly see that it was blue in colour with milky yellow patches inside, which from a distance had created the illusion of an orange colour. The ball floated for 20 seconds, passing the fridge and causing the motor to shudder, then travelling beneath a fluorescent tube and making this glow and finally traversing underneath a more ordinary light bulb causing the filament to glow. Then it curved back out through the window and disappeared.

This quiet mode of disappearance is not uncommon. The way ball lightning floats freely around a room is typical. The evidence that it was inducing radiated electrical energy into the gas in the light tube and electrical energy that was sufficient to heat up the filament in the light bulb must give important clues about the physics of this phenomenon. But, sadly, the Katong case was reported to a UFO group, not scientists. Despite many more examples like it, ball lightning is still not readily accepted by much of the scientific fraternity.

Victims of the thunderbolt

In August 1975 at Smethwick, West Midlands, a housewife stood by her cooker and suddenly noticed a small blue ball appear above her head. It then shot straight at her and hit her in the midriff with a muffled bang and a burning sensation. Apart from shock she was perfectly alright after her ordeal, but there were interesting physical symptoms that resulted. For example, the ball had burned a small hole right through her dress and panty-hose but had not damaged her leg at all. But she did suffer a limited numbness and tingling. Her wedding ring had a red burn underneath it. This suggested to physicists that eddy currents had been induced into the metal and, trapped in a continuous loop, had given off their energy as heat which had burnt the woman's finger.

This same effect has been reported in two classic UFO cases where there is more than a passing resemblance to the above incident. The

One of the few alleged photographs of ball lightning. Controversy surrounds the case, some alleging that this is a film fault and even a hoax. But it illustrates eyewitness accounts of what this mysterious phenomenon looks like

Authorities

There are few authorities on ball lightning, but the following works form a useful reference.

Two general articles by physicist Dr Paul Davies offer a good introduction in *Nature* 15 April 1976 and *New Scientist* 24 December 1987.

In the same issue of *Nature* Mark Stenhoff reports on the Smethwick, West Midlands, case with major physical effects. A further forensic report on the clothing from this incident was added by E. R. Wooding of Holloway College in *Nature* on 29 July 1976.

Books on ball lightning are rare and usually very complex unless you have a grounding in physics. Those worth referring to include:

■ Barry, Dr J., *Ball Lightning and Bead Lightning*, (Plenum, New York, 1980)

■ Randles, Jenny and Hough, Peter, *Spontaneous Human Combustion*, (Bantam, London and Berkeley, New York, 1993). Chapter 15 provides a layman's overview of the scientific evidence as part of an enquiry into its possible relevance to spontaneous human combustion (*see* page 92).

■ Singer, Dr Stanley, *The Nature of Ball Lightning*, (Plenum, New York, 1971)

only real difference is that in the UFO events the object was larger and lasted slightly longer. Therefore, it might have been a particularly virulent form of the same phenomenon.

Yet the woman attacked in her kitchen is considered a victim of ball lightning and the case has been widely discussed in the scientific press. The two UFO cases where eddy currents in wedding rings were reported have been summarily rejected by science as nonsense. Sadly, it is more than probable that by doing so physicists interested in what is potentially a rather dangerous natural phenomenon – ball lightning – are simply ignoring the better examples of their own subject. They do so simply because the witness has interpreted the event as a UFO encounter. This prejudice must be overcome.

In one example a ball fell into a rainwater barrel causing the rapid evaporation of the water and allowing energy calculations to be made. Other examples of 'hostile' ball lightning show that it is a violent phenomenon full of pent-up energy. In one case from Florida, USA, in August 1965, a woman on a veranda was using a swatter to brush flies away when she swatted a ball of light. The metal bat melted in the resultant explosion. Her friend nearby wrily noted, 'You sure got him that time!'

Do it yourself ball lightning

We now know a fair bit about the physics of ball lightning. Consequently we expect science to have resolved what causes it to form. But this is not the case. Perhaps the fact that it has only quite recently become widely accepted has limited the research possibilities. It took a case where a ball floated along the aisle of an Eastern Airlines aircraft in front of dozens of captive witnesses high above New York to really convince some researchers of its existence. One passenger was a top physicist. After that there was little recourse but to accept that something odd was happening.

But what? A major disadvantage is that there is almost no attested photographic evidence of a ball-lightning incident. Some events have been intensely debated and a camcorder film taken during a thunderstorm in Kent recently provoked some excitement until it was concluded that the ball that appears on the film was produced by the camera itself and was not an unprecedented glimpse of ball lightning in action.

This dearth of solid evidence is oddly akin to the problems faced by UFO experts. Possibly, if the two fields pooled their resources they might find that this elusive evidence does exist but has been recorded as UFOs rather than ball lightning. Faced with this problem, physicists have had to try to reconcile what are often real contradictions. Some cases suggest enormous energy is involved and others that there is only a fraction of that found within ordinary lightning bolts, for example.

As a result no theory of ball lightning formation has yet been widely accepted. There are those who favour an electrical discharge concept, but this struggles to fit in with those cases that form within a sealed space known as a 'Faraday cage' such as that formed by an aircraft hull. These are named after the physicist Faraday and provide a totally insulated and closed system where, theoretically, electrical activity should not penetrate. Others propose a nuclear radiation theory but detractors have tested locations struck by ball lightning and found no residual radiation. They also point out that close proximity witnesses should report symptoms of radiation sickness but nobody has – although some UFO witnesses have done so.

A particularly popular view has been that ball lightning is a form of atmospheric plasma. Japanese researchers such as Tokio Kikuchi and Yoshi Hiko Ohtsuki have paid particular attention to this. They believe this short-lived and relatively unstable state of matter trapped within a high energy field is rather rare in nature. They have, therefore, become very interested in the crop circles (*see* page 28) because fellow physicist Dr Terence Meaden quite independently proposed a plasma vortex theory to account for these events.

Some fascinating experiments have been conducted in Japanese laboratories where a very brief plasma vortex akin to ball lightning has been artificially created through high energy physics. The race is on to try to produce a plasma that is as stable as ball lightning appears to be and can drift freely for up to several minutes. The reason this is considered important is that these physicists suspect it might be the key to controlled nuclear reactions. The long-talked-about dream of cheap, unlimited energy by tapping the power of starlight might then come true. Science has sought this for decades without success and it may be that to understand ball lightning and successfully reproduce it in the laboratory would unlock the door, if, indeed, the plasma theory proves to be the correct one.

There is a growing mood of pessimism brought on by the failure to find a coherent theory and to reveal substantial hard evidence. In June 1988 Mark Stenhoff suggested in a paper for the *Journal of Meteorology* that the usefulness of eyewitness testimony, unless recorded within 24 hours of the event, was now open to serious question. In *The Skeptic* in 1992, the science writer, Steuart Campbell issued a challenge that the 'null' hypothesis – that ball lightning simply does not exist – should be taken seriously once again.

In the meantime, witnesses will continue to experience this strange phenomenon, often misperceiving it as a UFO sighting and scientists will continue to consider only but a fraction of the evidence.

Flashes of thought

- Ball lightning is a tennis- or basket-ball sized glowing mass that has a lifetime of between several seconds and a few minutes. It can appear suddenly in the open or inside buildings and floats close to ground level, sometimes buzzing or fizzing and disappearing violently with an explosion of light and energy.

- Despite its name it does not only occur during thunderstorms; although it is often reported in such conditions.

- Before the 1960s the scattered reports were treated as 'paranormal' by many scientists. Indeed some physicists still deny its existence today due to its rarity, transient nature and relative lack of good photographic or other solid evidence.

- Despite its recent legitimization, science has yet to find an adequate explanation or mechanism by which ball lightning can form. Attempts to recreate it in the laboratory have had only limited success.

- There is a strong possibility that many extreme cases of 'super ball lightning' are never reported because their nature is unrecognized by a witness. They may reach UFO groups. Ufologists misconstrue the nature of this data and physicists, who reject all UFO data, fail to take it into account. As such, theories about ball lightning may be relying upon incomplete, inadequate and possibly misleading evidence.

- The most comprehensive theory of ball lightning suggests it is formed from natural nuclear reactions. There are, however, no cases known to science where a witness in close proximity to the phenomenon exhibits evidence of radiation sickness, as the theory predicts they should. Several well-reported UFO close encounters do offer this missing evidence. These may, if accommodated by science, resolve the mystery of how ball lightning forms.

UNIDENTIFIED FLYING OBJECTS

Flying saucery

There have been UFOs as long as human beings have stared in awe at the sky. The reason is simple. The letters UFO, quite literally, describe anything which appears to be airborne and yet cannot be identified by the witness. It is not synonymous with alien spaceship or an exotic craft piloted by little green men although this is often assumed when the matter is reported by the media. This is compounded by investigators who fail to recognize that their primary task is to try to identify the unidentified. It is not to prove the existence of extraterrestrials. That is another thing altogether.

To ancient man much that was visible in the atmosphere was thought to be mysterious. The stars were regarded as gods and the constellations drawn around them. Even something as simple as a rainbow must have been baffling and frightening. Gradually the human mind would conceive that there were patterns within a rainbow's appearance. These happened when the sun shone after a rainstorm. Man began to appreciate that rainbows were a natural atmospheric phenomenon and not a mystical sign.

As the centuries passed rarer phenomena, such as ice haloes, were eventually changed from UFOs into IFOs (i.e. identified flying objects). Today puzzling things are still visible in the sky and we apply our own magic or mythology in order to explain them. The Greeks had their gods and goddesses to whom they could ascribe responsibility and, in the Middle Ages, wonders of nature were often related to the fairy folk but today in the space age we look beyond the Earth. Yet behind all of this sublimation is a reality – there are things within our atmosphere for which we still seek explanation.

While you may be unconvinced that there is a direct correlation between rainbows and UFOs there are good grounds for arguing this case. For instance, strange balls of light that float around, explode and can cause serious physical damage are well recorded. Without a knowledge of physics we would term them UFOs and either accept them as real or dismiss a witness as somehow befuddled. However, with the aid of physics we can now identify these things as ball lightning (*see* page 47), even if we do not yet know what ball lightning is.

UFO is a relative term and the long process of turning strange phenomena, to which we give romantic, mysterious names, into mundane phenomena, which we then teach our children in school science classes, is ongoing. We do not know all that there is to know about the world around us and the recent example of the acceptance of ball lightning is certain not to be the last.

A century ago not even the greatest mind on Earth knew about atomic reactions. Much of what we know as the twentieth century ends (from radio waves to genetic engineering) was not even dreamt of by science-fiction writers at the close of the Victorian era. Just as certainly there will be a science of the twenty-first and twenty-second centuries which is bound to take for granted things that none of us can glimpse today.

In such a context it becomes far less ridiculous to suggest that many of today's UFOs will prove to be examples of natural phenomena which we will one day comprehend. Scientists are wrestling with several things which may explain whole groups of UFO reports. These involve the transients or earthlights proposed by Persinger and Devereux from their independent research into geophysical processes (*see* page 34) and also the plasma vortex, invented as a concept by physicist Dr Terence Meaden in an attempt to explain crop circles, but which may be far more valuable in explaining UFOs (*see* page 28).

Both these ideas have developed from theory into laboratory research in the past two decades. Each idea has led to the recreation of earthlights and vortices within the laboratory. It is unlikely that they will be the last phenomena to be isolated. Other wonders are bound to be waiting to be discovered.

From airships to starships

Phenomena such as plasma vortices are not new but simply our awareness of them. They have been around since the Earth began and, as they cause UFO reports today, must have done so in the distant past. Indeed there is ample evidence of this. Greek and Roman literature is full of

A typical flying saucer, photographed in 1967 over a Madrid suburb. It represents the dream of alien spacecraft believed by many to relate to UFOs, but, as with most such spectacular images, was found to be a hoax

references to 'fiery shields' in the sky. Medieval tapestries and woodcuts depict globes of light in the sky with astonished citizens staring upward in wonderment. Most of these are probably examples of what today we would call a UFO.

In 1896 there was a wave of well-documented sightings of phantom lights soaring through the skies above California. The following year they spread to the mid-Western states. By 1909 and 1912 there were similar waves in New Zealand and Britain. Indeed in October 1912 Winston Churchill (then with the Admiralty) became the first politician known to have raised the subject of UFOs in front of any parliament.

Churchill did not, of course, call these silent lights UFOs or assume that they were the space craft of extraterrestrials. Although even in 1896 there were isolated examples where that solution was proposed. Instead, the culture of the turn of the century was obsessed with mankind's quest to develop powered flight. It was a dream about to be fulfilled but not yet fully realized. By 1896 Jules Verne, the Steven Spielberg of his day, was enthralling millions with his stories of what it would be like to fly. As a result these waves of UFO sightings were interpreted as mystery airships piloted by brave inventors who had finally cracked the much-anticipated secret. Churchill warned Britain that the lights over Kent could be craft developed by the Kaiser's troops who were preparing for a German invasion of Britain. Neither of these options was correct. The lights, whatever else they were, were not real airships. In every sense of the word they were UFOs.

The term flying saucer evolved in June 1947 when a pilot looking for a missing aircraft over the Rainier Mountains of Washington State, USA, told of his sighting of a formation of crescent-shaped objects. He thought they were secret aircraft. The US Army denied this and the media became very excited about the report, especially when other witnesses came forward to say they had seen things too.

Left Kenneth Arnold, the first modern UFO witness
Right Arnold's original book about UFOs

Intriguingly, that first witness, Kenneth Arnold, never said the objects looked like saucers. His sketches clearly show very different shapes. His words were that they sailed through the air like 'saucers skipping across water' (or as we might throw a flat stone across the sea and watch it bounce). Yet an enterprising journalist, either with an eye for a headline or because he simply misunderstood, invented the words 'flying saucer' and that term stuck. Curiously, people then started to see saucer-shaped objects, clearly suggesting that this shape owed more to the vagaries of human perception and imagination than it did to reality. Otherwise the coincidence was quite fantastic.

In fact this same process was in operation during the 1896 wave of sightings. Here people simply saw glowing lights. Once the stories began to appear that these were airships then airship shapes were reported and accounts followed describing these in graphic terms.

We now know this is a natural human tendency. Today there are plenty of cases on record where strange looking lights were seen by witnesses and described by them as alien spacecraft when there is no question that the witnesses saw no such thing, only peculiar lights. For instance, just after 19.00 hours on 31 December 1978, police and fire brigades all over northern Europe were alerted by reports of a crashing aircraft. Witnesses commonly reported they had seen a long body with lighted windows pouring fire from its rear. Emergency services went onto full alert. But there was no air crash. The object was the spectacular re-entry of a booster rocket from a Soviet space probe burning up as it hit the atmosphere, many miles above the Earth. All that was visible was a chain of blazing debris but most witnesses subconsciously 'joined the dots' and assumed they were seeing a single object sporting lots of windows that was far lower down that it really was.

A detailed study of some two hundred reports by UFO experts, later found that few (including those by police officers and airline crews) accurately described what was present. Most made false assumptions. Witnesses were split between concluding that the object was a crashing aircraft and a cigar-shaped alien spaceship with windows. All said it was very low down.

So common is this problem that all UFO investigators need special training to enable them to determine what a witness thinks was in the sky and then to assess what the source for that sighting might really have been.

The IFO problem

Skilled ufologists know that almost all reported UFO sightings are amenable to some explanation. Their surprising figures suggest that at least nine out of every ten reports are cases of mistaken identity. Some figures go even higher than that, but countless studies all over the world have endorsed these damning statistics. IFOs can be in many forms, from simple things like the planet Venus catching people unawares when it becomes very bright in the evening or morning sky, to more modern problems such as the increasingly common rain of space debris re-entering the atmosphere in spectacular fashion.

Sometimes a good deal of lateral thinking is required to solve a case. An owl that had eaten diseased fungus and was flying about glowing through bioluminesence was the source of one group of sightings. In another, a bin liner caught in the wind provoked a major flap in the West Midlands when many witnesses saw it as a slow moving dark object and called the police. Freak winds can make objects airborne, as IFOs, which would not normally fly.

However, ufologists are adamant that even though so many of the sightings turn into IFOs after careful investigation, there remain a few puzzling cases that do not. There are about five hundred sightings reported in Britain every year and thousands all over the world.

Much work has been carried out by the Space Centre in Toulouse, France, using money awarded by the government to study UFO reports. The scientists were not seeking little green men but new phenomena on the fringes of scientific understanding. They found, for example, that the greater the clarity of the atmosphere the more likely a UFO sighting was to remain unidentified. If the unresolved cases were simply further examples of IFOs for which not enough data had been accumulated, as some critics allege, the opposite should have been the case.

There is a tendency to assume that the unexplained sightings are examples of alien spacecraft, when that is, in fact, just one extreme theory favoured by only some ufologists. There has been an introduction of a new term. Just as in 1955 the term 'flying saucer' became outmoded and was replaced by a US Air Force acronym (UFO) so ufologists now refer to the rest of the unexplained activity as UAP – this stands for *unidentified atmospheric phenomena* and better characterizes what most of the residual UFO sightings more probably represent.

Have the aliens landed?

Although as many as 99 per cent of all sighting reports can be categorized as either IFOs, UAP or hoaxes and fabrications (fewer than two out of every hundred in most statistical surveys), the big question is whether any of the few remaining cases reflect some sort of alien contact. There have been about 5000 sets of UFO photographs taken in the past century. Some are hoaxes created by various trick methods, now often easy to identify with modern space-

Most UFOs turn out to be misidentifications of mundane objects. This fleet of craft over the beach area of Salem, Massachusetts is, in fact, lights in the room from which the photograph was taken. They are reflected in the window to create an illusion

Authorities

There are many books about UFOs and most are of limited reliability. Peter Hough and Jenny Randles have written a broad introduction to the subject entitled *The Complete Book of UFOs* (Piatkus, London, 1994). Jenny Randles's book *UFOs and How to See Them* (Anaya, London; Sterling, New York, 1992) is a very visual one about the problems caused by IFOs and how a witness can try to recognize what may have triggered a sighting.

■ *Dr Edward Condon* was a physicist who worked on the Manhattan Project developing the atom bomb. In 1967 he was assigned by the US Congress to head a study team at the University of Colorado spending a large sum of taxpayers' money investigating UFOs over two years. A 1000-page report was the result: *Scientific study of Unidentified Flying Objects* (E. P. Dutton, New York, 1969). The dossier contained extensive research into some sixty cases and detailed reviews of photographic, radar and trace evidence by various scientists. Although Condon's conclusions, at the start of the book, argue that the study showed that UFOs were of no scientific merit, the body of the text contradicts this. One third of the cases remain unresolved and scientists often suggested that they had to be termed 'unidentified'. Several of the team, upset by the way they felt the final report was being directed, left and wrote their own alternative scientific appraisal of the same data, concluding exactly the opposite – *UFOs? Yes!* by David Saunders and Roger Hawkins (Signet, New York, 1968). The infamous *Condon Report* itself is a goldmine of information and several scientists have stated that they became involved in UFO research when they read its content and its conclusions.

■ *Dr J Allen Hynek* was an astronomer who was contracted by the US government to be official consultant to the Air Force in its twenty-two-year project to study UFOs. At first Hynek believed everything could be explained, then he saw that some data was very impressive. When the Air Force closed its research in 1969 he wrote several books, including his excellent overview *The UFO Experience* (Regnery, Chicago, 1973). Spielberg's movie *Close Encounters of the Third Kind* was loosely based on this book and adopted the classification scheme Hynek devised for it (hence the film's peculiar title). Hynek appeared in cinema previews explaining the realities behind the mystery and even had a small part in the film, as a scientist fulfilling his dream and meeting aliens. In 1973 he also launched the Center for UFO Studies which still operates today. The Center pursues scientific research into the phenomenon. Hynek was never convinced that UFOs were spacecraft. He speculated that they might teach us something about the limitations of physics and somehow be connected with quantum reality. Dr Hynek died in 1986.

■ *Dr Carl Sagan*, the noted cosmologist, is a UFO sceptic but in 1954 participated in a top-secret CIA study to find ways to defuse public interest in UFOs. Along with Dr Thornton Page, who was a scientist, he has compiled a very important book, *UFOs: A Scientific Debate* (Cornell University Press, 1972) which contains papers from an American Association for the Advancement of Sciences symposium on UFOs held in Boston in December 1969. This was only days after Condon's report had concluded that science could gain nothing from the subject. The book is a very well-balanced collection of both pro and anti statements and assessments of the evidence from scientists.

■ *Dr Jacques Vallée* is a French computer scientist who moved to the USA as a young researcher to work with Hynek. His early books were the first truly scientific assessments of the subject, notably *Anatomy of a Phenomenon* (Regnery, Chicago, 1965) and *Challenge to Science* (Regnery, Chicago, 1966). He adopted a sociological approach and was the model for the character Lacombe in Spielberg's *Close Encounters.* He then developed a radical view seeing UFOs as a control mechanism like a thermostat regulating human consciousness. After a long absence he has published a trilogy *Dimensions*, *Confrontations* and *Revelations* (Ballantine, New York, 1988, 1989 and 1991; Souvenir, London, 1989, 1990 and 1992).

■ *The New UFOlogist*, 71 Knight Avenue, Canterbury, Kent CT2 8PT

■ Center for UFO Studies, 2457 West Peterson Ave, Chicago, IL 60659, USA

■ UFO Research Australia, Box 2435, Cairns QLD 4870, Australia

age technology. However, most photographs turn out to be further examples of IFOs, because if the eye can be fooled then so can a camera. Indeed, at times the camera is more susceptible. For example, if a bird flies across the field of view the camera operator may well not see it. When the photograph is developed an image frozen onto the film by the fast shutter speed may often make the bird look like a strange disc-like object.

It is common for a UFO not to be seen until a photograph is developed and these are almost invariably shown to be a result of the photographic process, perhaps a drying mark or scratch on the negative that has coincidentally taken on UFO-like proportions. Investigators almost never take seriously any cases where the witness says they did not see the UFO that later turned up mysteriously on their film.

However, there are a few extraordinary photographs that appear to show objects rather than just amorphous lights. These are very rare and often date back many years. Despite the increase in camera ownership and sophistication of film the number of photographs in this category has actually dwindled, something that no ufologist has successfully explained. Yet it clearly must be a significant factor.

It is possible that these are all just very good hoaxes that have survived attempts so far to try to expose them but most ufologists refute that argument. Equally they might show examples of Earth technology such as secret aircraft under test.

There is mounting evidence that new generations of remarkable military aircraft have regularly been misperceived as UFOs before

Left A common type of IFO (identified flying object) – a rocket burning spectacularly as it is launched above Russia in 1968. *Below* A not dissimilar-looking 'UFO' photograph from South Carolina, USA, in 1971. This may also be a blazing mass of gases in the atmosphere rather than an alien craft

UFO facts

■ By definition (as aerial phenomena unidentified by observers) there have always been UFOs. Many have now been identified by science, as aurora, fireball meteors, meteorological processes, etc. These all seemed very mysterious to our ancestors, without the benefit of this knowledge.

■ Today, 95 per cent of all reported sightings of UFOs have a conventional explanation after investigation. Over 200 different objects have been mistaken (from aircraft lights to owls glowing in the dark). This still leaves several hundred puzzling sightings worldwide each year.

■ The modern era of UFOs began with sightings in June 1947 over the American Pacific West. The term 'flying saucer' was invented by a journalist who, in error, thought these things were said to be of a saucer shape. In fact they were not described as such by witnesses. However thousands have since claimed to see craft shaped like 'flying saucers'!

■ The US military was charged with a secret investigation from 1947 onwards. Afraid of being invaded by secret weapons from the USSR that study was made top secret. The idea of a 'cover up' to hide the 'truth' about 'alien invaders' has resulted from that starting point.

■ When proven that UFOs were not from any power on Earth the cultural dream of life in space in the early 1950s became a powerful influence on perceptions of UFOs. In fact, hard evidence that any UFOs are alien spacecraft is much less than many assume. It is far weaker than the more general and defensible case that some UFOs represent phenomena so far unexplained by science. Nevertheless, the idea that UFOs are synonymous with alien craft is very deeply engrained within our society.

■ There is almost certainly no single solution to the unsolved UFO cases. They represent a variety of new phenomena, with most, if not all, natural in origin, such as super ball lightning, earthlights, plasmas etc.

the existence of their technology is officially confirmed.

Reports of triangular UFOs in Belgium between 1989 and 1991 are widely believed by some investigators to have been operations by American stealth jets straight off the drawing boards, even though the flights seem not to have been known about by local air force commanders.

Similarly, an object which smashed a hole through pine trees in Rendlesham Forest, Suffolk, on 26 December 1980 left physical traces in its wake, doubling the background radiation. This was investigated by senior air force personnel from the NATO air bases of Bentwaters and Woodbridge, who seem to have been unaware what they were witnessing. A political impasse developed between NATO governments over this worrying affair. There are even strong reasons to think that speculation about the object being an alien spacecraft was allowed to grow, since witnesses could not identify what they saw. Indeed this idea may even have been fostered in the Pentagon in order to prevent the case becoming widely recognized.

A spaceship crash would be considered a wild tale fit only for the tabloid press. If the truth involved the use of some technology so secret that even the local base personnel were uninformed about it, then it may have been expedient to prevent that truth from becoming common knowledge.

There may well be military reasons why a subject is obscured and UFOs act as a perfect foil to hide things from the public. Of course, UFOs exist. All world governments accept that. The British Ministry of Defence, like that of every major nation, collates sightings and keeps records on file. It appears to do so on the basis that nobody can afford to take chances and discoveries may be made from the data collected, notably by the scientific staff to whom the evidence is passed.

There are strong signs in close-encounter cases that emitted radiation is occasionally involved, stopping car engines, burning witnesses' skins, etc. But none of that establishes an alien reality behind the phenomenon. As we have seen there could be several types of UAP that have associated energies but which have nothing to do with alien contact. If aliens are to be found then evidence for that lies elsewhere – in the extraordinary reports of the so-called abduction phenomenon.

ALIEN ABDUCTION

Space pirates

Alien abduction has been described variously as the greatest mystery of the space age, an inexplicable psychosis sweeping the world or the first contact with another intelligent lifeform. Nobody knows which is correct.

This remarkable phenomenon of space piracy has now been alleged by thousands of people from all cultures and walks of life. While many retain an objectivity about what has taken place, a number of abductees do insist that they have been used as human guinea pigs by scientists from another world. This is the sort of concept that is predictably unacceptable to all bar a very few people.

The phenomenon seems related to, and yet considerably post-dates, the UFO mystery. While sightings of strange lights in the sky can be traced back thousands of years, the alien abduction is very much a product of the space age. It began in the late 1950s but only came into its own during the 1980s. However, it is now so widespread and has such disturbing overtones that psychologists, psychiatrists, medical doctors, sociologists and UFO experts, particularly in the USA, have all joined forces in a concerted effort to unravel the truth. Are abductions a subjective experience like a dream or fantasy or are they in any sense objective intrusions into the lives of ordinary adults and children?

More than one such researcher has braved professional rebuke by supporting the suggestion that the true number of abductees may be huge – possibly a significant fraction of the Earth's populace. Such statements were widely dismissed as hysteria until hard evidence was published in early 1992. The Roper organization carried out a major survey in the USA and found that up to four million Americans could be victims of an alien kidnap. This equates to just under one million abductees in Britain – and as much as 2 per cent of the world's population.

Even more disturbing is the further allegation that very few of these people are consciously aware of what has happened to them. It is argued by some researchers with increasing conviction that anybody may have been the victim of an alien abduction without having any recollection of what has taken place.

History

In September 1961, Betty and Barney Hill, a social worker and a post-office employee respectively, were returning over night from a holiday in Canada. As they headed south through the rugged White Mountains towards their New England home they encountered a banana-shaped light in the sky. At one point it moved closer and they thought they saw beings looking out of windows at them. The couple made it home without further incident, or so they thought at the time.

In fact, there were strange blotches on the paintwork of the car, and they were inexplicably over an hour late upon arrival home. Not long after, the couple started to share bizarre nightmares about alien faces and medical examinations. Unable to handle the resulting stress they went to see Dr Benjamin Simon, a Boston psychiatrist, who began a long therapy which included the use of regression hypnosis.

Their dreams and phobias were traced back to that fateful car ride. While Dr Simon assumed that he was treating a delusional nightmare, not a real memory, he let the amazing story unfold from these therapy sessions. Eventually, both husband and wife had a coherent account of being stopped on the road by a 'spacecraft', being taken onboard it against their will by little men with oriental eyes and white faces, and then being given probing physical examinations.

Betty was frightened by one procedure which involved a needle inserted through her navel. The beings told her this connected with a pregnancy test. Several years later a not dissimilar procedure was devised for withdrawing samples of amniotic fluid from a woman. These could then be tested for Down's syndrome. Samples of hair and blood were reputedly also taken from both victims by the entities. At one stage they were even shown a 'map' which located the aliens' home world in terms of trade routes between stars.

The story caused a sensation when details were published in 1965 by writer John Fuller. Astronomer Marjorie Fish even took Betty Hill's star map sketches and reconstructed an interpretation of what they might mean. This required the use of information that could not

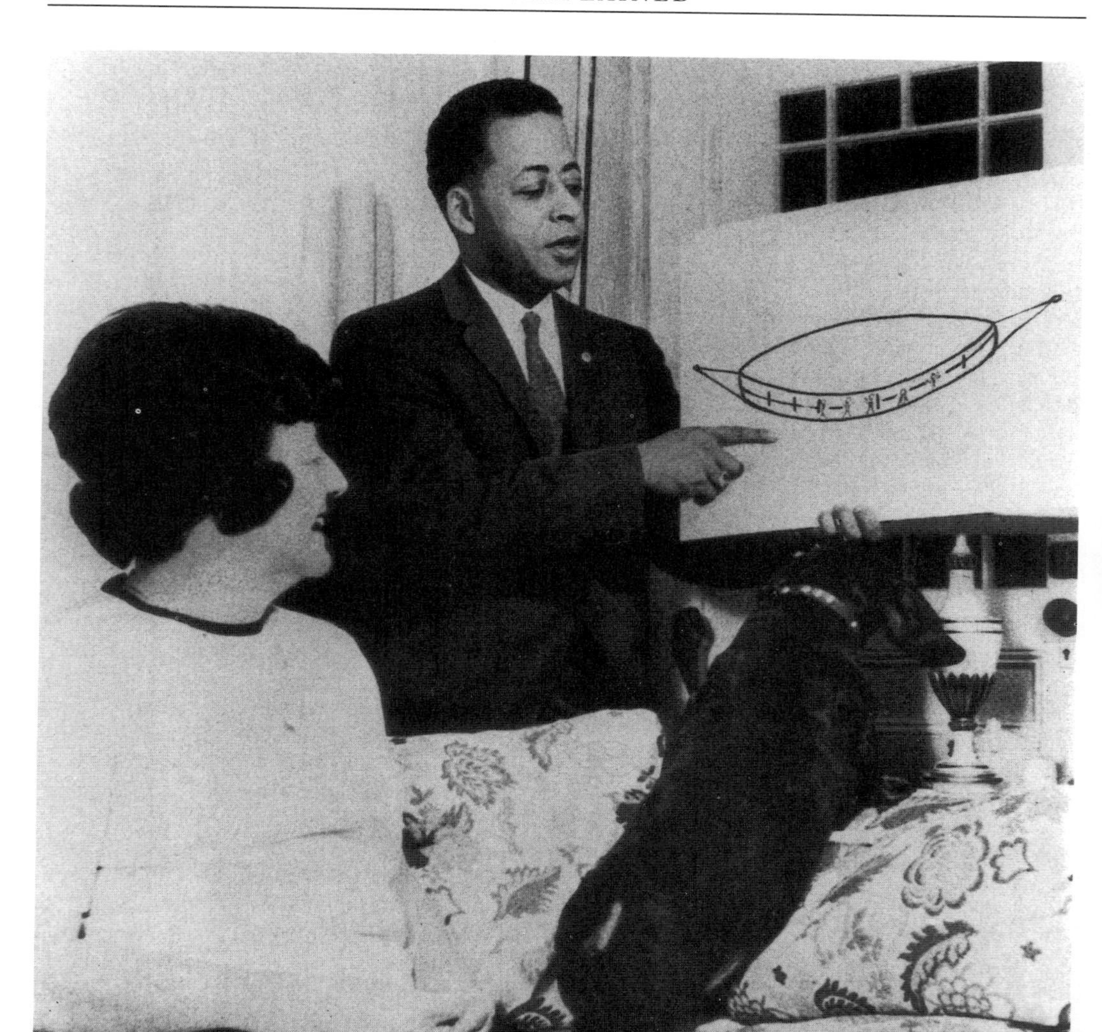

Betty and Barney Hill show a sketch of the craft and alien beings which they believe abducted them from the White Mountains, USA, in September 1961

have been known by Betty Hill in 1961 but was known some years later when Fish carried out her study. The outcome gave the location of the alien world as a planet around the star *zeta reticuli.*

The debate surrounding this case still rages in the UFO community. Sceptics alleged that the Hills' spaceship was an illusion based on the planet Jupiter then very bright in the same part of the sky. However, an object was recorded on radar at the time by a local air base. Sceptics also note how the psychiatrist felt that the witnesses believed what they were saying but had, in his view, suffered a shared hallucination. Even so the Hills did not exhibit any other indications of any sort of psychosis. Whatever the truth, this remarkable case opened the floodgates. In the long term, a torrent of abductions was set to follow.

Credible witnesses, incredible stories

By the early 1970s several psychiatrists in the USA were using regression hypnosis to take a witness back to a puzzling encounter whenever there was any indication of loss of or missing memory. Some witnesses were quite credible, even if the resulting stories were rather harder to accept.

Police patrolman Herb Schirmer was in his car on the flat country roads of Ashland, Nebraska, one night in December 1967 when he came upon a strange object that was apparently sucking power from the roadside electricity wires. Silvery suited creatures took him on board and examined him, asking questions and then explaining how their machine operated through magnetic fields.

An investigation was mounted by a team of scientists from the University of Colorado, using major funds awarded by the US government. They failed to prove he suffered from any mental aberration. Schirmer's police chief confirmed that he believed in the patrolman's sincerity and Dr Leo Sprinkle, the psychiatrist brought in by the university to assess the witness and conduct regression hypnosis, also came out in favour. Indeed Sprinkle went on to devote many years to the study of hundreds of other abduction cases.

A remarkably similar case befell West Yorkshire police officer, Alan Godfrey, who saw a rotating object hovering just above the road during an early-morning patrol outside the Pennine mill town of Todmorden in November 1980. Similarly, Godfrey only consciously recalled the sighting of the object and noticed that a few minutes of time had disappeared from his life during a later reconstruction. Until the use of regression hypnosis he had no story to plug the gap.

Godfrey was regressed by doctors from a prestigious Manchester practice on St John's Street and graphically relived his terror for a video camera used by researcher Harry Harris. A beam of light shot from the hovering object and rendered him unconscious. He then described himself being floated into the object and medically probed on a long table or bed by small robot-like creatures who seemed to be under the supervision of a taller 'humanoid' figure who was called 'Yosef'. A big black dog was in the room also.

Some years after the hypnosis Godfrey told us what he thought of the experience. 'I know what I saw on the road that night was real. As for what I later remembered under hypnosis, I

Former West Yorkshire police officer Alan Godfrey sketches the UFO and alien entity involved in his November 1980 spacenapping at Todmorden

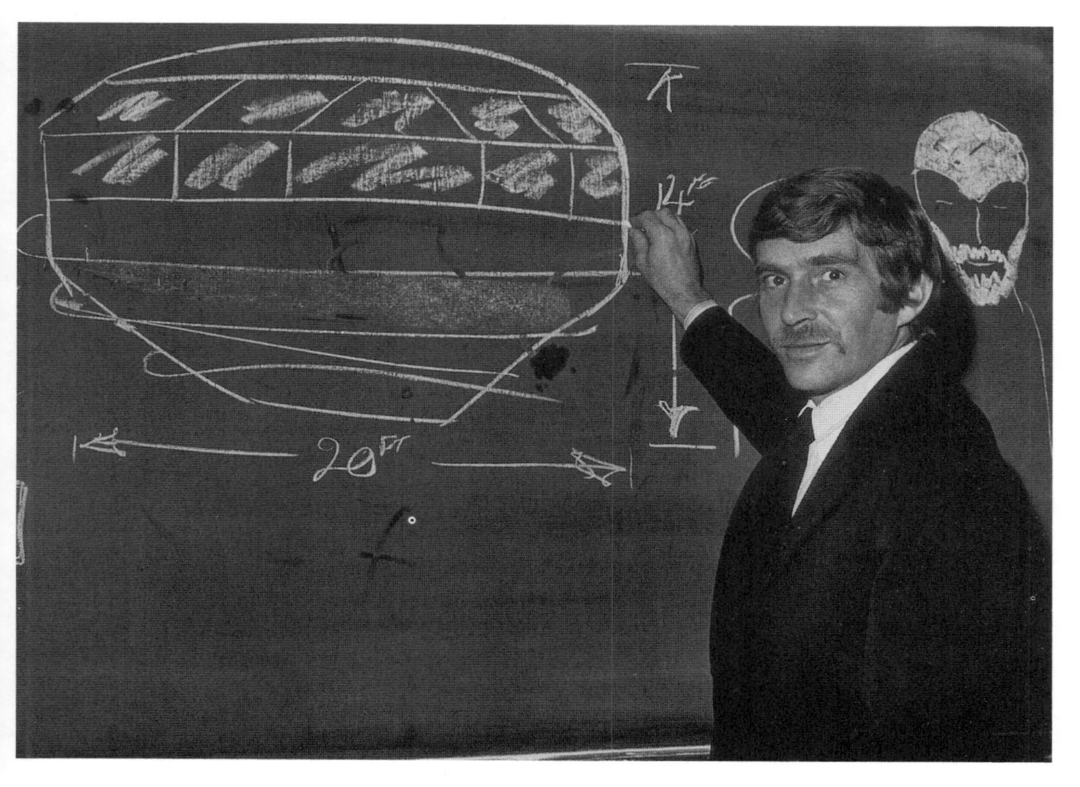

have no way of telling if I was describing a real memory or a fantasy based on something I had read.' However, there was one clue. His sturdy police boot was mysteriously split, as if he had been dragged along the ground.

The problem is that, while there are similarities between these early cases, the entities involved in each one are different, the things that happen to the witness while inside the UFO are not the same. Furthermore, there is an element of surreality in the storylines, such as Alan Godfrey's meeting with the big black dog.

You are our children

Britain's first abduction case was investigated in 1977, but occurred in October 1974. A family of five – two adults and three children – had been returning from a night at a relative's house when they encountered a weird bank of green mist that was stretched out across the road near their home at Aveley, Essex. They drove into it, felt a bump and emerged from the other side seemingly seconds later. In fact, when they reached home about half a mile later, over an hour and a half of their lives had simply vanished from their minds.

In the months to come the family went through a series of remarkable changes. They became vegetarian, took a great concern in the environment and found that they shared the same frightening dreams. In these dreams they were each seeing a small hairy creature with terrifying eyes.

Eventually, Dr Leonard Wilder conducted hypnosis on the two adults and they put together an incredibly detailed account of their experience for UFO experts Andy Collins and Barry King. The two witnesses told of being aboard a UFO which had apparently lurked behind the green smokescreen. They described how the entire car was 'beamed' aboard the object. They then experienced an out-of-body state where they floated around the interior of the object aware of their bodies still in the car. They were medically examined by small hairy creatures a bit like something out of the television programme *Dr Who* but which were supervised by taller, human-like beings wearing silvery suits.

Afterwards the father was given a tour of the 'engine room' and a long explanation about how the craft operated. Then they were shown holographic images of a devastated planet which the aliens claimed was their home world. They did not want Earth to suffer similar decay and intended to help us ensure this did not happen. Mankind is a genetic experiment placed on this world by them millenia ago and they have since maintained a constant watch over us. People were being abducted, they claimed, to facilitate monitoring of the population. They wished to change the world by stealth rather than direct intervention. It had to be our own decision to put things right.

All in the mind?

More cases like these began to flow in from all over the world. By 1987 there were about five hundred on record, with over 50 per cent from the USA. Despite consistent details in general outline the sceptics were quick to reject the evidence. They focused on the very real problems of regression hypnosis.

Experiments have shown that hypnosis can stimulate memory in a subject, for example, who they sat next to in the classroom as a child, but just as easily it is capable of triggering a fantasy. Under regression people are known to be much more open to cues and suggestion from the hypnotist and have in some tests been shown to develop a totally false memory of an event that never happened merely because a hypnotist suggested that it had during a regression experiment.

In one case a patient was led to believe they had heard a gunshot after being subtly coaxed into this under hypnosis. When brought out they insisted that they had heard the shot. Then they were played the tape of their original interview hours before in which they had adamantly insisted they had heard nothing.

It was evidence such as this that caused regression testimony from witnesses to an accident or criminal act to be largely excluded from the judicial system after initial optimism that this work might prove very beneficial.

It was proposed by researchers such as aviation journalist Philip Klass that abduction stories were emerging because of the expectation of both investigators and witnesses. They believed in UFOs, thought they had seen one, assumed UFOs were alien and so, when led into this 'territory' by subtle suggestion, concluded that they had met their crew. After the Betty and Barney Hill story spread around the world, first in the Fuller book and in 1975 in a TV movie called *The UFO Incident*, it was conjectured that nobody was immune to knowledge of the for-

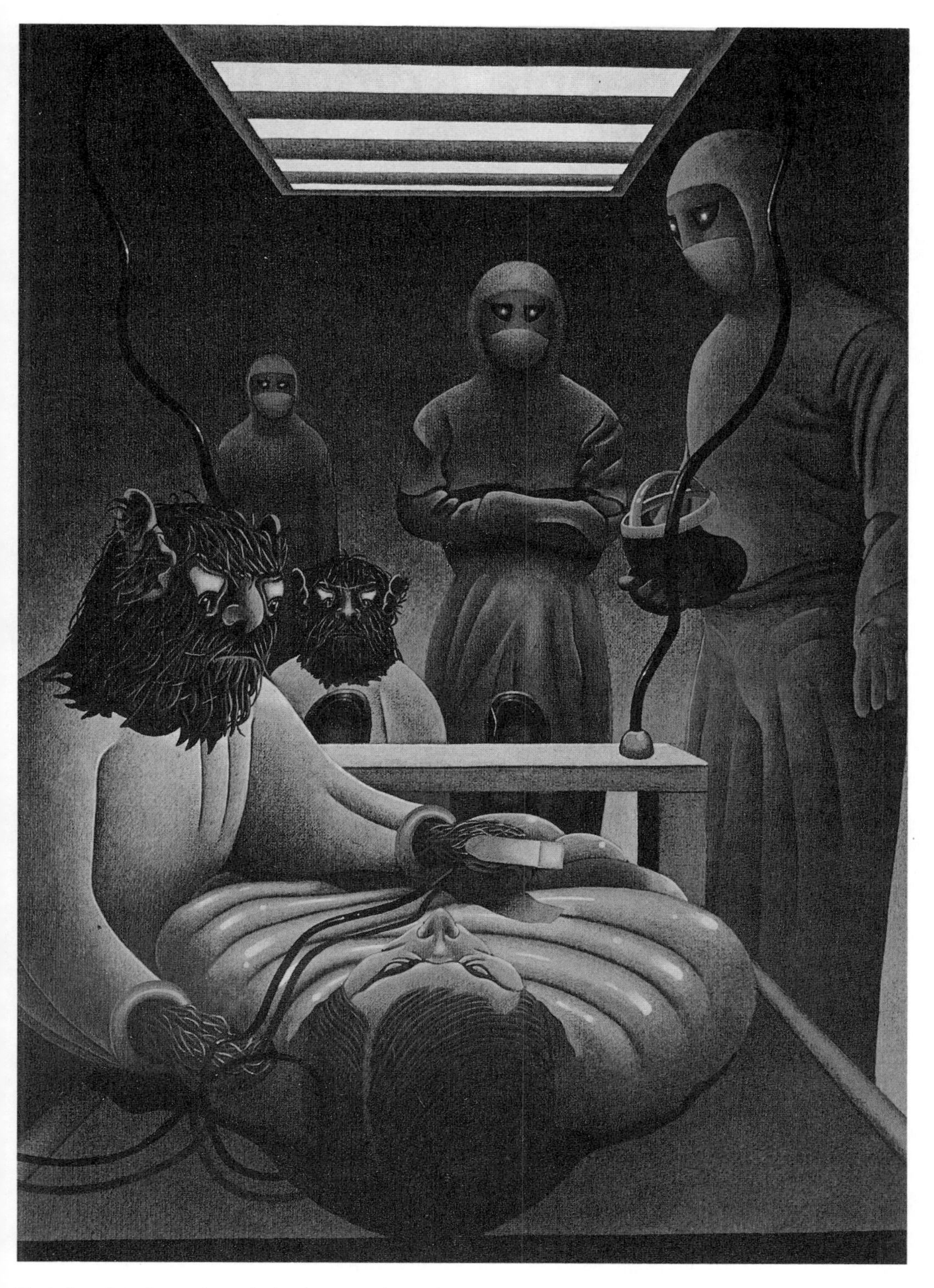

Most alien abductions rely on single witness testimony, but in an encounter at Aveley, Essex, in October 1974, an entire family of two adults and three children appear to have had a missing-time experience

mat that an abduction might take. The subconscious mind was readily able to do the rest and create a fantasy in that special state of consciousness that was regression hypnosis.

In order to demonstrate this Alvin Lawson, a Californian professor of English, appealed for student volunteers who knew nothing about UFOs. He asked them to take part in regression hypnosis experiments. These were carried out in conjunction with Dr William McCall. A series of leading questions were chosen to fit the framework of real abductions. The subjects were then told to fantasize an abduction story. Only eight students took part but the experiment was later widely cited by sceptical scientists as proof that all abduction cases could be imaginative in origin.

In fact the experiment left itself open to serious criticism. Inevitably the fictional accounts were similar to the real ones because the fantasies were closely structured by the questions that were asked and these emerged from supposedly real abductions. Even so, there was a wide diversity of aliens reported by the students, several of which had no precedent in genuine case histories. Equally, there was no emotion expressed by the students, and none of them were left with even a remote belief that their encounter had happened.

In spontaneously reported abduction cases quite the opposite is true. Emotional response is often very severe, so much so that one witness suffered an epileptic seizure during regression and doctors have had to terminate other sessions when heart rate and blood pressure reached dangerous levels. There is undoubtedly a strong conviction during the process that a witness is reliving a real experience.

Several detailed profiles of abductees have been carried out by psychologists. In at least one case, the psychologist worked blind in that she was merely asked to comment on patterns within a group of people she believed she was vetting for vocational purposes. All these studies have revealed that the witnesses are of above average intelligence and display no trace of psychosis. Even after being told the truth about her sample of people, the psychologist was unable to revise her opinion.

Indeed some research by psychologists has noted that the closest parallel emotionally is with victims of rape. Abductees genuinely feel physically and mentally abused regardless of the reality behind the accounts.

Space-age fairy tales?

Dr Thomas Bullard is a folklore researcher from the University of Indiana. He became interested in abduction stories when he realized that they might be modern folklore in the making. Bullard launched a huge research project to try to prove this idea and this statistically correlated over four hundred well-researched cases. In 1992 he updated this study to include almost eight hundred.

Some significant facts immediately emerged. About 40 per cent of the cases came without any recourse to regression hypnosis and there was no detectable difference between the data which did arrive this way and that which did not. As a result he was forced to conclude that, while the use of regression techniques might add some fantasy to the details of a story and could compromise individual cases, the facts strongly support the view that the alien abduction phenomenon is not simply a product of the hypnotic state. This is merely a cloudy glass through which a very real phenomenon is viewed.

Bullard also noted an odd feature of cases that he called 'doorway amnesia'. In science-fiction movies, the moment when the victim is taken onboard the spacecraft by alien captors is often given great prominence. The relevant scene is very dramatic and all those with a vivid imagination sense the importance of displaying flashing lights and sliding doors, as in movies such as *Invaders from Mars* and Steven Spielberg's UFO inspired *Close Encounters of the Third Kind.* However, in all real abductions this moment of entry into the UFO does not occur. The witness has no memory, either consciously or later under hypnosis, of precisely how they went inside the UFO. They simply 'jump' to a point where they awoke in a strange room.

In order to test this and to confirm or deny the Lawson/McCall imaginary abductee experiment, Jenny Randles used a non-hypnosis technique on twenty subjects in Britain in early 1987. They were asked to fantasize an abduction just as the students had been in California, but now the questions were open-ended, asking only for the subject to describe being taken into a UFO and then relating in detail what happened.

Ten of the subjects were knowledgable about UFO studies. The other ten only had a superficial knowledge gleaned from newspapers or television. The results were startling. The

Dr Thomas Bullard, a folklorist from the University of Indiana, has made a major study of alien abductions and attempted to show that they are space-age mythology

subjects aware of UFOs described cases very like real abductions (often with doorway amnesia and entities similar to those that are encountered). This was not unexpected. What was surprising, however, was that the subjects with only a little knowledge were very different. Most described exactly how they entered the UFO and referred to methods, motives and types of alien beings completely unlike those encountered in real alien abductions. In fact they were very much in the mould of science-fiction movies.

Sex monsters from outer space

The most disturbing aspect of the alien abduction evidence may possibly be the consistency that weaves these cases together. Reports have been gathered from all over the world, although almost 50 per cent are still from the USA, where research is particularly obsessive. Yet even in countries where there has been limited publicity of the subject, such as in the former USSR, the accounts seem very like those that are found elsewhere. There is also a clear historical trend worldwide that can be traced throughout the evidence.

In October 1957 Antonio Villas Boas, a Brazilian farmhand, was allegedly taken into a UFO and rubbed down with a chemical before being raped by a flame-haired female alien yelping like a dog. Afterwards the being pointed to her belly and then up at the sky as if gesturing that she was to have his baby on another world. The case was investigated by a medical team soon afterwards and kept quiet for years as too fantastic – even to report to ufologists. But it was documented before Betty Hill told of her pregnancy testing during the White Mountains abduction. Indeed Barney Hill described a technique used on him which later was recognized

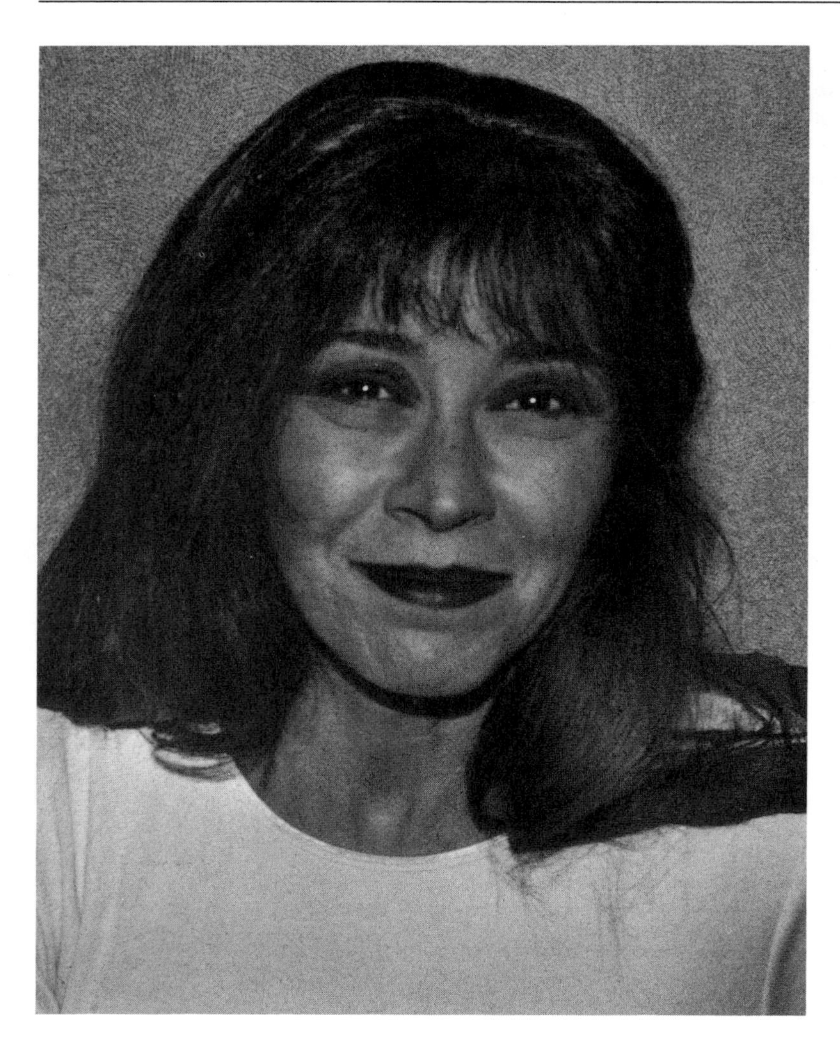

The ultimate encounter is believed by many to have befallen this woman. Linda Napolitano claims to have been abducted from her high-rise apartment block in Manhattan – an event witnessed by others including, allegedly, a world-renowned statesman, said by some to be a former UN Secretary General

as the then unknown medical extraction of semen. At that time the Villas Boas case had not even been published in Brazil and so was independent corroboration.

Throughout the years, elements of genetic sampling, mating and sexual reproduction have dominated abduction cases. In an obscure but intriguing case from Venezuela in August 1965 a highly respected gynaecologist was contacted and told by tall, blond-haired blue-eyed 'humanoid' creatures that, 'We are here to study the psyches of humans to adapt them to our species. We are studying the possibility of inter-breeding with you to create a new species.'

In May 1968 a young woman from New York State had to have one year's intensive treatment from a gynaecologist when she stopped menstruating after an experience where she had vividly recalled ova samples taken from her and strange beings explaining that they were trying to make a hybrid baby from human and alien cells.

In 1973, before news of this case could possibly have reached her, a woman in Somerset, England, told of similar samples being taken and of later being raped by a tall 'humanoid'. She felt sexually abused.

Wise babies

In the mid 1980s Budd Hopkins, a noted New York artist and abduction researcher, started to publish data that he had collated for years but had thought nobody would believe. This described the extraordinary claims made by certain young women.

Hopkins' victims said that they were abducted, raped and later re-abducted, during which time the now-developing foetus was extracted from their womb and their pregnancy terminated. Some claimed later they were shown a successful hybrid baby 'grown' by the aliens in their own environment and that this was super-intelligent. Many of his female abductees then started to report what became known as 'wise baby dreams', which they initially rejected as fantasies.

Remarkably, while Hopkins was debating whether to publish this startling new development in the evidence, similar material was emerging around the world. A woman from Cheshire, England, had an encounter in the summer of 1979, and on 16 September reported a dream in which she saw herself bearing a strange-looking and extremely intelligent baby.

On 26 December the now pregnant woman at the focus of this case suffered an unexpected and complete disappearance of her foetus. At Pudasjarvi in Finland in April 1980 another woman was kidnapped, examined and told candidly by her captors that they needed her help because, 'we cannot beget children of our own.'

In abduction cases from 1980 onwards there is a strong bias towards young, female victims – the age range eighteen to thirty-five being predominant. Over 65 per cent of witnesses fall into this group. In three cases where the abductees were past child-bearing age their report indicates a clear element of rejection. One male witness was told by the abductors, 'You are too old and infirm for our purposes.'

In the USA today a major medical search is underway to come up with solid evidence that could prove these missing foetus cases once and for all. So far the physical proof is completely lacking, but several doctors are convinced that the circumstantial evidence is so strong that it can only be a matter of time until the abduction phenomenon is no longer in dispute.

Abductions at M.I.T.

The renowned Massachusetts Institute of Technology sits across the Charles River from Boston. In June 1992 it played host to (but did not organize) the most important gathering of alien abduction researchers that the world has yet seen. With financial backing, the facilities of a world-famous university and the scruples and standards of a traditional scientific symposium, five very full days were devoted to over a hundred detailed presentations by the specially invited participants. These were psychiatrists, social workers, professional abduction investigators and the largest collection of abductees ever gathered together for such a purpose. The intention was to assess every single aspect of the mystery, from how it was first discovered to methods of helping a victim to cope with its aftermath. This was carried out in a controlled environment, free of the hype or publicity that UFO conferences attract.

The published proceedings of the event were designed to become a comprehensive manual for the growing number of professionals paying attention to the field. Jenny Randles was one of the delegates fortunate enough to be invited.

Psychiatrists produced paper after paper reviewing results of studies into the psychological profile of abductees. All these tests established that witnesses were overwhelmingly stable, sincere, and had no detectable delusionary or psychopathic disorders.

Several tests were conducted into the so-called 'fantasy prone' hypothesis, proposed by some psychologists. This argued that abductees could be part of a small group of humanity (up to 5 per cent of the population) who had such rich inner fantasy lives that they grew up with a significant difficulty in distinguishing real events from imaginary ones. Each of three independent tests negated the hypothesis and Australian researcher Keith Basterfield, one of the key proponents of the theory, formally withdrew it from the debate.

Specialists in the field of post-traumatic stress syndrome also showed how abductees had clear profiles which matched those of victims involved in other real cases of physical abuse. This supported the claim that these people believed in the reality of their abduction encounters.

Abductions from the USA could be compared with studies carried out in Britain, Australia, Brazil and Scandinavia. Jenny Randles produced a report which highlighted new themes emerging from British data. Two key facts this data isolated was that the abductees were found to be visually creative and to have extraordinary recall of their very early lives, as early as one year old or less. These may prove vital clues.

What are abductions?

Many people take abduction stories at face value and assume that they must represent alien

Authorities

■ *Dr Thomas Bullard* is a folklore researcher from the University of Indiana who has compiled two extremely detailed theses which provide a massive statistical breakdown of hundreds of abductions. He attempted to prove that they were modern folklore but has failed. He is uncommitted to any solution but feels that the extraterrestrial theory cannot be eliminated as yet. His work has been published through the UFO community and is greatly respected worldwide.

■ *Budd Hopkins* is often considered the leader in abduction research. A New York artist, he has investigated reports since 1975 and now performs his own regression hypnosis. He has investigated hundreds of cases and believes that an alien genetic experimentation programme may be underway with not altogether beneficial results for the victims. His work was dramatized in the four-hour American TV mini-series 'Intruders' (CBS, 1992). His first book was *Missing Time* (Marek, New York, 1982) and was followed by *Intruders* (Random House, New York,1987) which tells the story of one woman's repeated abductions in Indiana. Hopkins lectures widely around the world.

■ *Dr David Jacobs* is a professor of history from Temple University who began to investigate abductions in the mid 1980s and believes that he has isolated a precise medical procedure which indicates the nature and purpose of an alien mission to Earth. His thesis appears in the book *Secret Life* (Simon and Schuster, New York, 1992; Fourth Estate, London, 1993).

■ *Leo Sprinkle* is a psychologist called in to investigate an abduction in 1967 by a US government-funded team from Colorado University. He has investigated over two hundred cases since then. Sprinkle believes that abductees may represent a contact with non-human intelligences that are largely benevolent towards us. He has published no books but each summer organizes the 'Rocky Mountain Retreat' at which abductees come together to share their experiences.

Other useful works in this field follow a very different line to the normal American approach which accepts the alien visitation hypothesis as virtual fact. These are as follows:

■ Basterfield, Keith, *UFOs: The Image Hypothesis* (Reed, Sydney 1980) – an Australian perspective of the psychological approach.

■ Evans, Hilary, *Gods, Spirits, Cosmic Guardians* (Aquarian, Northants, 1986) – a psycho-social approach to the problem, proposing a unique form of hallucination as the solution.

■ Klass, Philip, *UFO Abductions: A Dangerous Game* (Prometheus, Buffalo 1989) – a sceptical examination accusing researchers of distorting the evidence and playing hazardous games with the minds of witnesses.

■ Randles, Jenny, *Abduction*, (Robert Hale, London, 1988; Headline, London 1989; also a US edition retitled *Alien Abductions*, Global Communications, New Jersey, 1990) – a global review of all the evidence, patterns and theories for abduction cases. An updated sequel, *Aliens: The Real Story* has been published (Robert Hale, London, 1993).

■ Ring, Dr Kenneth, *The Omega Project*, (William Morrow, New York, 1992) – takes a very different look at abductions from a psychologist who specializes in the near-death experience. He describes a major statistical survey of both abductees and near-death experiments, concluding that they share a great deal and may both represent the next phase of evolution of the human mind towards a state he calls Omega.

■ Strieber, Whitley, *Communion*, (William Morrow, New York, 1987; Century, London, 1988; and several foreign-language translations) – the famous horror writer's own account about finding a hidden pattern of abductions within his life. Intriguing and unusual speculation about its nature.

Horror writer Whitley Strieber, whose terrifying true-life alien abduction claims in New York state made a huge impact on the world when published in his 1987 book *Communion*

contact. But this is by no means a proven fact and there are many who seek alternative solutions.

The quest for a psychological explanation of some sort has not been lessened by the finding that abductees are apparently normal people. All of us can have extraordinary mental experiences without being in any sense unstable or ill. But no mechanism has yet been found which can explain the abduction to the satisfaction of the psychologists who have researched its evidence. However, there are some pointers that suggest how the experience seems to be created within the mind. For example, there is a social pattern to the 'aliens' which emerged particularly in cases reported before 1987.

The year 1987 was a watershed because Whitley Strieber, the horror novelist, after reading about the abduction mystery in a book by Jenny Randles, realized that odd dream-like experiences in his past could turn out to be real memories. He worked to uncover them and reported in two autobiographical books called *Communion* and *Transformation*. Millions of copies were sold worldwide and *Communion* was made into a big movie staring Christopher Walken. From that point onward the entire world knew intimate details of the abduction phenomenon, and particularly of one specific description of an entity known by researchers as 'the greys' – which as the name suggests has a grey skin and is small in stature with an oversized head and large eyes. Since 1987 this has gone on to dominate all abduction accounts throughout the world.

Yet before Strieber's books appeared, the 'aliens' seemed to have differed markedly in their physical appearances in various parts of the world. In South America hirsute dwarves with aggressive behaviour were more normal. Only in the USA were the abductors small egg-head scientists. Britain and northern Europe tended to attract tall, blond-haired creatures looking like Norse gods and with rather more civilized manners.

Why was this? Some researchers suggested that there were many alien races visiting Earth. Others argued that the true origin was our own future and the forms reported mere disguises adopted by time travellers to hide the truth and perhaps, therefore, sharing some element of national trends from today. Others propose that there is much too close a match between the culture of a nation and the form that its abduction cases adopt. There must be a psychological reason why the entities have been perceived in this changing manner. The theory was that after Strieber's case became well known it set a new template.

If an alien race is in contact with us, our minds are probably incapable of conceiving and handling that fact. Studies of primitive island or tribal cultures show that when they first met 'civilized' man they perceived us according to images that were culturally appropriate to them – for example, the old adage about seeing an aircraft as a 'big white bird'. Their minds simply had no other way of coping with information their brains had never been programmed to contain. Perhaps we similarly perceive aliens according to our own expectations and clothe the unknown in science-fictional imagery.

This latter view was at least partly endorsed by Dr Richard Gregory, a noted perceptual psychologist. In *New Scientist* on 30 August 1962, long before the abduction debate beset his colleagues, he had already warned that, 'Suppose we were to meet something really odd – say a new life-form . . . could we "see" it properly?'

The search for proof

Budd Hopkins, the abduction researcher, presented a remarkable report on a case which seems to be the world's most evidential alien encounter. It occurred on 30 November 1989 in Manhattan, New York where a woman claims that she was abducted from her bedroom by several small entities. After being floated through the air high above the streets she entered a spaceship and was medically probed. The witness, Linda Napolitano, had a conscious recall of this event and reported it to Hopkins, who explored it further under hypnosis. The woman had already reported several experiences to him earlier in 1989 after she read one of his books. Hopkins and this woman were, therefore, working together on a professional basis when her major encounter took place.

Then Hopkins was contacted by two law enforcement bodyguards who said they had been outside in the streets in the middle of that night when they saw the woman floating out of her high-rise apartment and into the ship. Without knowing that Hopkins had already been contacted by Linda they told him that they could pinpoint her home and lead him to her. The witness accounts matched very closely.

Further confirmation from other allegedly independent observers on a bridge over the East river has since been received in dramatic fashion by Hopkins. He is continuing to investigate this case as he feels it is potentially the most important in the history of alien abductions. But there are problems. He has been unable to interview face to face the two police bodyguards who later advised that they were taking a senior 'political figure associated with the United Nations' to the New York heliport and that the whole event may have been staged for that person's benefit. It is argued that were a person of this stature to admit this observation, alien presence would be proven and the cover-up ended. Some researchers claim the man was former UN Secretary General, Javier Perez de Cuellar, but this has not been confirmed by either de Cuellar or Hopkins.

Sadly, the heliport deny there were any flights that night. The 'world leader's' office insist he was at home in bed. Attempts to find other witnesses in the apartment block or a busy newspaper office directly opposite have all produced nothing, according to an independent study published by another investigation team. Thus the most promising of cases has failed to produce much hard evidence to support itself.

However, it has a rival in a case first reported directly to us less than two days after its occurrence and investigated in great depth by Peter Hough. On 1 December 1987 a former police officer claims that he was walking across Ilkley Moor, West Yorkshire, just before dawn in order to take some landscape pictures, when he saw a small green creature gesturing at him. He ran after it in time to see a disc-shaped craft streak

Abduction abstracts

■ *Earliest known case* Some examples from 1942 are known but they have all been reported since 1975. There are strong correlations between modern alien abduction accounts and stories over several centuries of individuals abducted by fairies and taken inside 'hills' and on trips to 'fairy land', where time runs on a different scale.

■ *First recorded case* Antonio Villas Boas in October 1957, Brazil.

■ *First widely published case* Betty and Barney Hill, White Mountains, New Hampshire, USA, September 1961, but published, 1965.

■ *Total number of cases* Dr Thomas Bullard has a catalogue of eight hundred well-researched cases. True number reported as of September 1992 believed to be about 3000. This is widely conjectured to be just a few of a far larger number, with estimates ranging up to many millions.

■ *Most examples* The USA has more abductions per head of population by far, followed by Britain, Brazil, Argentina and Scandinavia. Australia has remarkably few reported cases. Non-Western cultures such as China and India, have almost none. This is a significant omission rarely addressed by abduction researchers given the enormous size of the indigenous population of these countries. These figures, however, might not reflect the true number of cases. They depend on active investigators in any country, and the likelihood of people reporting such an experience. This can be shaped by social attitudes.

■ Some 65 per cent of cases involve some use of hypnotic regression. The British UFO Research Association imposed a moratorium on this technique for five years after 1987. It did not prevent a few conscious abductions being reported which differed little from earlier cases.

■ *Witness profile* The average abductee is aged twenty-seven, and over 60 per cent are female. The average number of witnesses per case is 1.20, indicating that they are very often single witness events.

■ *Event profile* The abduction follows a remarkably consistent pattern. First the witness sees a bright light or entity, followed by a loss of awareness of time and space (known as the Oz factor). Then they find themselves inside an unknown room where strange beings perform a medical examination. Next, information is often given to the victim and images may be shown of a home world or coming cataclysm. The witness is then given promises about future visits or a role they must fulfil before finding themselves back in the real world with full, partial or no conscious memory at all of what has just occurred. A fuller memory can happen spontaneously over weeks, even years, or a short cut can be attained through hypnotic regression. The cases tend to follow this structure so exactly that folklorist Bullard says it disproves the folk-tale narrative theory. Folklore presents a far more fluctuating order and pattern.

skywards, but not before he was able to take one colour photograph of the being (see p. 164).

The case has physical evidence in the form of a compass needle which (the witness alleged) had reversed its polarity. This and the photograph were analysed by professional research bodies such as a Manchester University and the Kodak laboratories. The witness also suffered a reputed time-lapse of about one and three-quarter hours and was regressed by a clinical psychologist, Jim Singleton after the case had been thoroughly explored in the hope of discovering more information. A full-scale abduction memory emerged from this and Singleton confirmed that he believed the witness to be recounting a real experience.

This amazing case is the world's first abduction where the victim successfully photographed his captor. But the photograph can only show a 4½ foot (1.37 metres) tall greenish figure. No analysis can establish whether that is a dummy, a child in a suit or a real alien.

OUT OF TIME

SYNCHRONICITY

Just a coincidence

In February 1987 a woman from Warrington, Cheshire, was being plagued by the telephone. It was ringing at all hours of day and night, usually with nobody on the other end when she picked up the receiver. After several such frustrating experiences the telephone rang again and she shouted out in sheer exasperation 'shut up!' Instantly it did so. After just one ring the telephone was silenced and it behaved normally ever after.

This is what most of us would describe as a coincidence. But how many would feel it necessary to assume that there was a causal link between the woman shouting out and the telephone stopping as she did? In her mind there was a connection, even though it was unclear what that link might be. 'It was as if the phone heard me and took notice,' she explained.

Of course, if we analyse the situation, what seemed to this woman to be a strange event was not really terribly strange at all. Most of us get annoyed if the telephone rings at an inopportune moment. Some of us no doubt put our anger into words. Equally there are certainly times when the phone starts to ring and then stops again, usually because of a fault on the line, misdialing by the party at the other end, or perhaps because they were interrupted as they made the connection and stopped it before getting all the way through.

We can work out the odds for each of these events. We might shout at the telephone once every hundred calls. The telephone may stop suddenly about as often. From these figures we can reliably suggest that about once every 10,000 calls there would be a coming together of these two circumstances in such a way that something paranormal appears to have happened. When taking into account the huge number of telephone calls made around the world each day this 'amazing coincidence' will turn out to be a regular occurrence and a surprise only to this woman because it was her turn that day.

Long odds are always coming up. One out of a few million people will win a lottery, all the rest will not. During a game of cards there have been several well-attested cases where four players have been dealt four whole suits at random from the shuffled pack. The odds are indeed phenomenal but given that so many hands of cards are dealt each day they are simply bound to come up from time to time.

Under these circumstances it is difficult to see why anybody should assume that there is anything supernatural about the wonders of chance or coincidence. Indeed some psychologists reasonably point out that this is a perfectly adequate explanation for so-called precognitive dreams. Since trillions of dreams are dreamt every single night it is statistically certain that some of these will occasionally seem to match an event that later befalls the dreamer. Human imagination then takes over making the dream more similar to the real event through the process of selective memory and a mystery is manufactured. John Grant in his book *Dreamers* (Ashgrove, Devon, 1984) notes that we never hear about the billion dreams that do not come true, only the one that does.

However, is this solution absolutely convincing? It can be tested. In 1929 J. W. Dunne reported in his *An Experiment with Time* how easy it is to make a 'dream diary' recording one's dreams and then comparing them in detail with those events that follow later. Many researchers have done this using the more sophisticated technique of speaking dreams into a tape recorder. In this way figures suggest that about one in every hundred dreams contain what seems to the dreamer to be a preview of the future.

Jenny Randles recorded an example of this in 1968 when she saw in a dream a paper factory by a bridge that was on fire. Less than twenty-four hours after the dream was recorded she went on a university sponsored walk from Lancaster to Manchester and at 04.00 hours she was walking with a group of people through Preston, when she saw an orange fire painting the sky. They rushed to the scene, with one voice exclaiming it was a real fire. When they reached the building, right beside a bridge over the River Ribble, it turned out to be a paper factory engaged in a night shift. There was no fire but the error was made. Jenny had no memory of the dream, only a vague sense of *déjà vu*. Upon arrival home she checked her diary and was astonished at the comparison. The matching of details seemed extraordinarily exact.

Synchronicity

Can dreams such as these be a trick of chance or a truly strange phenomenon? As in most seemingly paranormal events this revolves around a minor incident of trivial importance, as indeed do many images within dreams themselves. But where can we draw the line between what is an acceptable match by way of the statistical argument and what goes too far beyond the limits of chance? How often before, or since, would one dream about either a bridge (perhaps quite often?), a fire (less often, but probably once or twice) or a paper factory (the strangest of the three key components to this story)? However,what really made it seem paranormal was the way in which all three elements combined together and matched a real event only hours after the dream. This combination would probably have been almost (but not quite) impossible just by guesswork.

Interestingly, research has shown that these things are more likely to occur when an out-of-the-ordinary period in one's life is about to occur. Here the coming sponsored walk was an exciting thing in the life of a teenager. It may have provided the link that forged this strange incident within the subconscious mind. This was either a very unusual coincidence or a precognition, if indeed there is any distinction between the two.

Dr Carl Jung, the renowned psychologist, and Dr Wolfgang Pauli, a quantum physicist, came together to consider these problems and produced the theory of synchronicity. This attempted to show how seemingly unrelated events could link together at the level of consciousness in some sort of unseen relationship. Jung felt something like the concept was needed to explain amazing coincidences of the precognitive type. Pauli knew that in sub-atomic physics there was an increasing view that everything that happened emerged out of a random chaos and that the order that we do see is somehow tied in with the consciousness of the observer.

At first the theory was controversial. In recent years experiments have been carried out to try to explain why certain sub-atomic reactions happen as they do. The astonishing results that have emerged from this work imply that if someone observes its outcome then an experiment will go in a certain direction, but without intervention from this determining factor the result remains in a haze of probability or chaos.

Types of coincidence

Researchers have isolated different kinds of coincidence. The one we are mostly familiar with is called the 'bunch', a group of seemingly random events which stress a particular theme. This appears to give some credence to the old wives' tale that things come in threes. Indeed, once some sort of incident occurs during the bunch type of coincidence others like it seem more likely to follow. Part of the reason here may simply be that we pay more attention to a subject once it comes to the fore and so notice examples of it more acutely than we might otherwise do.

Another type of coincidence is much stranger and seems to have more of a paranormal flavour to it. This is what is called the 'idea reinforcement'. There are a number of examples in science where a discovery has been made randomly and simultaneously by several different people. This is not strange when theoretical research is following an identical pattern in various laboratories and the same outcome is therefore bound to occur given duplicated circumstances. But when the planet Neptune was discovered, for example, at least two (possibly three) teams of researchers using different methods appear to have found it after decades of fruitless effort at more or less the same moment.

There almost seems to be a ripple in the collective unconscious of mankind which ensures that when an idea's time has come it is responded to by several different people at once.

Perhaps even stranger still is the motivational coincidence. Here there seem strong grounds to suspect that what appears to be a coincidence on the surface may in fact result from driving forces at the subconscious level which have effectively manipulated the person to be in the right place at the right time. It is as if they were aware of it at this level but that sentience was not translated into conscious knowledge. Instead the mind manoeuvred them into a series of actions which led them instinctively into a position to take advantage of some dimly perceived subconscious truth. The result was what to all intents and purposes seemed to be an amazing coincidence.

The Jenkins family had been raised in Yorkshire, then moved south to Leicestershire, but in August 1983 decided to visit their daughter who had moved to Edinburgh. Forty years before, the father had served with a man called McKirdy from that city while with the RAF in Egypt. On the spur of the moment, and knowing nobody else in town, Mr Jenkins decided to look up his friend.

Of course, he had no idea if that man was alive or dead or had long since moved from the city. Nor did he know how many McKirdys there might be to search through. They found only one in the telephone book but after some deliberation, fearing embarrassment by calling the wrong person, Mr and Mrs Jenkins decided instead to go on a bus tour.

They arrived at the coach station and selected a particular trip and waited, but it was cancelled because too few customers arrived to take that excursion. Mrs Jenkins now decided to call her daughter and explain, so they sought a telephone, finding one eventually by the taxi rank. As she telephoned, Mr Jenkins waited outside staring at the road. Moments later a taxi pulled to a halt right beside him bearing the driver's name, as is the unusual custom in this city. That name was McKirdy and its driver was indeed his old friend from the RAF.

The interesting point here is that this coincidence involved a number of behavioural decisions by Mr Jenkins. He chose not to call the number in the book (had he done so he would have got no reply and probably given up). He and his wife chose to visit the coach station,then to select a tour which quite unusually was cancelled. Then they searched for a telephone and found one by the taxi rank. Mr Jenkins chose to let his wife make the call and to stand in the only spot where they would have then been able to read the name on any taxi that came to a stop.

As you can see, although this seems like a random event it depended to a considerable degree on Mr Jenkins making a series of choices. If, somehow, his subconscious mind was aware that his friend was a taxi driver and would be coming to a halt by the coach station soon afterwards it could have subconsciously motivated his choices so that he was in the right place at the right time to undergo this meeting.

If you feel that this may be difficult to believe, consider another case where subconscious motivation is more strongly implied.

In July 1980 Charlotte Richards was in her kitchen at Tacoma, Washington in the USA with her mind in an idling state – something that seems very significant as a trigger to the coincidental events. It is as if the subconscious awareness can sneak across the barrier much more freely if there is not a lot of other input clouding the mind at the time.

As she washed the dishes Charlotte's husband was mowing the lawn outside. It was a typically idyllic domestic scene. But suddenly her subconscious kicked into motion and a mood of depression swept over her. In the back of her mind she saw an image of blood and knew that this meant something. Interestingly, her rational mind deduced (as anyone's might) that this was a warning of an accident to her husband. She had an urge to shout out to him but she did not do this. Her subconscious mind seems to have seized control and made her body react in exactly the opposite way to which logic decrees it should have done. Instead of stepping up to the window, banging on it or calling out, she stepped back from the window and ducked away.

It was fortunate Charlotte Richards did this. For a moment later, by a million to one fluke, the lawn mower hit a stone, threw this into the air like a rocket and hit the window full on with enormous force. The glass shattered into tiny fragments, showering the sink. Had she not stepped away moments earlier there is little doubt that this woman would have been horribly injured.

The only way we seem able to resolve a case like this is to suppose that Charlotte was motivated to act as she did by her subconscious mind which had accessed momentary foreknowledge of the events that were about to take place. Some of this was glimpsed consciously as a brief vision.

If it can happen this way when someone is idling as we all may do when washing up the dishes, it can happen elsewhere. However, perhaps Mr Jenkins was too busy with other conscious sensory input to be aware of the 'little voice' telling him what to do, or feeding images into his conscious mind, so instead his subconscious had to manipulate a coincidence by guiding him to take certain random actions that led to a remarkable event.

Make your mind up time

The role of coincidence in behavioural matters is little understood as few scientists are following Jung's and Pauli's lead and exploring the subject. But it is well illustrated by a remarkable chain of events that befell Jenny Randles in May 1983. All of them are well documented.

At the time she was presenting a feature for Radio City in Liverpool in which she discussed a different aspect of the supernatural each week. At the conclusion of each specially written ten-minute feature she chose the subject for the following week's programme. She then visited the studios to record a promotional piece that would be aired on the station to plug that show. Finally she would come in and do the new programme and the cycle would begin again.

That week Jenny decided she would discuss synchronicity and joked that it was an adage of this subject that when you wrote about coincidences these started to happen in your life. Afterwards the producer, Wally Scott, expressed some concern to her that such an obscure topic was to be discussed and described by an obscure word. 'Can't you call it something else?' he asked. 'Don't forget this is an audience of mostly teenage, record-buying people. They are not likely to know a word like "synchronicity".'

Anyhow, the item was forgotten for a few days until it was time to write the promotion for the new show. That weekend a strange thing happened. There was an armed robbery nearby and the thieves dumped their getaway car right outside Jenny's front door in Wallasey. This was not the smartest move they might have made as a Merseyside CID officer lived directly opposite! Jenny spent some time assisting the police with their enquiries as a result.

Jenny prepared to go into Liverpool to record the trailer with this fresh in mind but before she left she took a telephone call routed through from the radio telescope at Jodrell Bank. This happened from time to time if someone reported a UFO sighting. On this occasion someone in a district of Merseyside had seen a 'star fall out of the sky', as he phrased it, which had occurred very early that morning.

So with this consigned to memory Jenny went into Liverpool and upon arrival at the station learnt that there had been a major tragedy that morning. Two youths had without obvious explanation abandoned their motorcycles and fled onto a railway track in the middle of the preceding night. They were hit by a high-speed freight train and killed. Police were expressing puzzlement as to why this had happened.

Jenny was even more baffled. She checked all

Author Jenny Randles, who has researched various types of synchronicity after her own extraordinary experience

Coincident conclusions

■ Some events do occur by nothing more than random chance and are what we would call a pure coincidence. The odds that two people at a party will share the same birth date are surprisingly low. But if that coincidence happens to you it seems supernatural.

■ However, there are some coincidences which appear to defy all the statistical odds and serve a real function in the life of an individual. Some psychologists and many paranormal researchers believe these are important to teach us things about mind, consciousness and the perception of reality.

■ Psychologist, Carl Jung, and quantum physicist, Wolfgang Pauli, distinguished a category of coincidence which they called 'meaningful' and which seemed to be created by some purposeful agent acting for a reason, albeit in obscure ways. They termed this 'synchronicity'.

■ At the level of sub-atomic matter all things are determined by statistics and coincidence. There is a debate in physics as to what 'hidden variable' exists to cause one solution to emerge above any other. A major candidate is that the agency responsible is human consciousness.

■ Meaningful coincidences could occur as a result of manipulated behaviour patterns and indirect influence over quantum mechanical reactions. These would occur to engineer at a subconscious level chance events that serve a specific purpose. Order would be created out of chaos and an 'amazing coincidence' would result. In truth it would simply be a consequence of the way in which reality is structured.

the incoming reports on the Radio City news-desk. A gut feeling was confirmed and, indeed, the incident seemed to have occurred at exactly the time and in precisely the same district where the man who had recently telephoned claimed to see a star fall out of the sky. It was at least a possibility that the youths had seen the same object, left their bikes and chased after the falling star (which seems most likely to have been an unusually bright meteor) and as a result lost their lives by unwittingly stumbling onto the railway line.

The question was what should Jenny do now? Should she risk implicating the man who had innocently reported the falling star? Should she look a fool by telling her story to the police? Indeed was it right to intrude on the grief of the family of the two victims by adding news of a story that they might consider ridiculous or even sensationalist?

Still torn by this, Jenny recorded her trailer promoting the 'synchronicity' feature and went into the studio to talk to Brian Ford. He was the radio presenter she worked with on her item and he also had a daily show. They sat talking while the music played as Jenny sought advice on what to do.

At that time there was a huge glass window separating the studio from the Radio City reception area so that anyone entering here could watch the disc jockies in action and, of course, those in the studio could see out into reception. At the very moment Jenny asked Brian Ford if she should notify the police, in walked two uniformed policemen off the street. In fact, they had come to interview a member of staff about a motoring offence which subsequently made a news story in its own right. However, the juxtaposition of this with the question racking Jenny's brain at that precise moment was quite extraordinary.

Amazed and now more than half persuaded that she had been given her answer, Jenny left the studios to return home. On the way she was handed a free newspaper detailing forthcoming record releases. Normally she would not have read it but as there was a long bus journey through the Mersey tunnel she glanced at it to pass the time. Her astonishment was absolutely confounded by a story about a new album tipped for the top that coming month – the album by world-famous band, Police, was almost unbelievably to be called *Synchronicity*.

A few weeks later the album and the title track single both went to number one in the charts. The lyrics, penned much earlier by Sting, are indeed about Jung and Pauli's theory of coincidence and contain words which on the face of it would seem rather baffling to what the Radio City producer had called a 'teenage, record-buying audience'. Yet in the light of the above story the words appear utterly incredible. One line says, 'A star fall, a phone call, it joins all – synchronicity.'

As you might expect, Jenny Randles did tell the police, but there is no startling ending to this story. The police said the information passed on was as useful as anything else, but it did not solve the case. So what did this sequence of incredible events mean?

A whole series of incidents seemed to link Jenny Randles with the police and synchronicity in a way that stretches credulity and yet has no obvious motivation. Clearly she was persuaded to do something she might not have done before this chain of events. But for what reason? We may never know. But this does suggest that there are deeper levels to human consciousness than seems obvious at first glance.

As mentioned at the start of this report, thinking about this subject often seems to stimulate events. Did it do so here? Perhaps. The above piece was written in April 1993 on the day BBC television screened an episode of a popular science-fiction series called 'Quantum Leap'. The main character is, like Pauli, a quantum physicist. Each week he is thrust into a different month between 1950 and 1999. That particular episode by chance concerned the physicist helping police with their enquiries and was set synchronistically in May 1983 the exact month of the above story.

Of course, you may prefer to accept that there is simply no meaning. All events are random. But there does come a point where a pattern goes beyond what is acceptable to the rational mind and so we impose outside order. Perhaps synchronicity is just that.

A SLIP IN TIME

Is it possible to move through time from the present moment either forwards or backwards? Can we visit the past and observe life as our ancestors experienced it, or leap forward to a place populated by our children's children?

There are two fundamental theories which attempt to explain what time is. In the first, time is a linear phenomenon where only 'now' exists. What is past is finished with and cannot be revisited. The future does not yet exist, and when it does it becomes the present. Linear time is readily and easily embraced by most of us, based as it is on chronological time. Yet it is man who devised the calendar and invented the wristwatch to bring some order out of the chaos of living. If this hypothesis is correct then time travel is an impossibility. Yet time struggles to escape this man-made confinement. Our timetabled lives buckle under the pressure of days which seem to drag, and other days that contain too few hours.

In the second theory, all time exists at once. The past, present and future become relative terms dependent on where in time 'we' are presently located. J. W. Dunne wrote a book in the 1920s called *An Experiment With Time.* It contained Dunne's analysis of his dreams recorded over a long period. They were frequently precognitive, predicting personal events in his life and events of global significance. He saw consciousness as the route to a true understanding of time. His dreams were a mixture of past, present and future events, demonstrating, he concluded that all time existed at once.

Time exists in a continuous loop. If we accept this model then it might be possible to journey across the loop to a point either in the 'past', or in the 'future'. The whole of time could be likened to the images on a spool of video tape. All that would be required to conquer time travel would be the knowledge of how to operate the fast forward and reverse controls.

This may sound fantastic but we already know that time is not as mundane as suggested in the linear hypothesis. Albert Einstein in his theory of relativity mathematically demonstrated that to an observer travelling close to the speed of light something very strange would happen. Such an observer, for instance, leaving on a journey in a spacecraft and returning fourteen years later, would discover that sixty-five years had passed on Earth. A twenty-two-year trip would mean that almost a thousand had elapsed back home. In effect, he would have travelled into the future. Time slows down for an observer travelling at very high velocity. Even stranger, our hypothetical space traveller would shrink in size, and decades would pass between meals as measured back to Earth. None of this would matter – onboard the ship everything would seem normal and

German physicist, Albert Einstein, developed his theory of relativity at the beginning of the 20th century. It sought to explore and define the interdependence between matter, time and space

time would flow, relatively speaking, at its customary rate.

If time, sometimes referred to as the fourth dimension, really is such a nebulous concept, then our lack of understanding of it means that justifiably anything is possible.

A step backwards

Pensioner Mrs Charlotte Warburton who lived with her husband near Tunbrige Wells in Kent, glimpsed backwards through time on Tuesday, 18 June 1968. The couple had travelled into town to shop, then went their separate ways, planning to meet up later for a coffee at a regular venue.

After making her regular purchases, Mrs Warburton tried several shops for a tin of shortcake. During the search she went into a small self-service store which was unfamiliar to her. There was no shortcake, but while she was looking around the shop, the pensioner noticed an opening in the left-hand wall and curiosity drove her to investigate. It was the entrance to a large mahogany-panelled rectangular room, in stark contrast to the modern chrome and plastic fitments of the shop. Mrs Warburton commented, 'It had no windows that I could see, but was illuminated by a number of electric bulbs with small frosted glass shades. I saw two couples in mid-century dress, and can quite clearly remember the outfit of one of the women. It consisted of a beige felt hat tilted at an angle and trimmed on the left-hand side with a wisp of dark fur. Her coat was also in beige, and was quite fashionable though rather long for 1968.'

She also noticed several men in lounge suits and a glass-enclosed cash desk nearby. Everyone was drinking coffee and chatting, which did not seem out of place considering it was mid-morning. However, it seemed odd that she had not heard of the coffee shop before, and later realized that she could not *smell* any coffee. When she met up with her husband, Mrs Warburton told him of her discovery and they agreed to try the new coffee shop the following Tuesday.

A week later they did their shopping as usual then entered the small store and walked to the place where the entrance to the coffee shop had been. Where the entrance had been, a frozen food display cabinet now stood against the wall. Adamant she had not made a mistake, Mr Warburton accompanied his wife into two other similar shops, but to no avail. The experience was so vivid she now wondered if her perceptions had momentarily slipped backwards to a period when the mahogany-panelled room had existed. Charlotte Warburton decided to do some research.

She contacted a local woman interested in psychic matters, and asked her if she remembered the building. Up until several years before, she was told, a cinema had stood next to the shop. Over that, and to the left of the shop was the Tunbridge Wells Constitutional Club. The woman remembered visiting the club during World War II, and it had small tables for refreshments and mahogany-panelled walls.

Not content with that, Mrs Warburton traced the present location of the club and contacted its steward, who had been in the position since 1919. He informed her that the entrance to the old club had been via a street door and a flight of stairs to the left of the store. On the second floor was a refreshment room which tallied exactly with Mrs Warburton's description.

There are many cases where people have apparently glimpsed places as they were in a former time. Ruth Manning Saunders records an incident on Dartmoor in her 1951 book *The River Dart.* Three girls accompanied their father on a shooting expedition at Hayford, near Buckfastleigh. During the afternoon the girls wandered off and as darkness fell, became lost. Then, according to the author: 'To their joy they saw a light ahead and found a roadside cottage. Ruddy firelight danced out from uncurtained windows, warming the night with a friendly glow. The three girls looked through the window and saw an old man and woman sitting crouched over the fire. But, on a sudden, lo, the fire, the old man and woman, and the entire cottage vanished, and night, like a black bag, fell over the place.'

Buildings from another time were also apparently seen by the American biologist and paranormal investigator, Ivan T. Sanderson. He, his wife and an assistant were driving in a remote area of Haiti when their car became bogged down in a pool of mud. They abandoned the vehicle and began to walk until exhausted. In his book *More Things,* Sanderson wrote: 'Suddenly on looking up from the dusty ground I perceived in the now brilliant moonlight, and casting shadows appropriate to their positions, three-storeyed houses of various shapes and sizes lining both sides of the road.'

The scene developed further as the ground became cobblestoned and muddy. His wife pointed and in shock described the same scene. Sanderson was convinced the buildings were Parisian. After marvelling at them for some time the couple began to feel very dizzy. Sanderson called out to the assistant who was some distance ahead. The man turned back and the biologist begged him for a cigarette. As the flame from the assistant's lighter was extinguished, so was the scene from fifteenth-century France. Furthermore, the assistant had been oblivious of the vision and had noticed nothing unusual.

Author Joan Forman experienced her own time-slip while researching her book on the subject, *The Mask of Time.* She was visiting Haddon Hall in Derbyshire during a break from work. While pausing in the courtyard she 'saw' four children playing at the top of some steps. The eldest, a girl of about nine, had her back to the author. Ms Forman described the child's white Dutch style hat, long green-grey silk dress with lace collar, and shoulder-length blonde hair. She could hear their laughter, yet was aware that her perception of the children was not through the physical apparatus of the eye. Suddenly the girl turned to face her. She had imagined a face of beauty, but the girl was very plain indeed. Momentarily shocked, Ms Forman took a step forwards and immediately the children vanished.

She went inside the Hall searching for a portrait of the girl and eventually found one. The child depicted in the painting was younger but their was no mistaking the broad jawline and snub nose. Joan Forman had apparently witnessed Lady Grace Manners at play, many years after her death.

Haddon Hall in Derbyshire. When author Joan Forman visited the house, she caught a glimpse of a former time. While standing in a certain spot, she 'saw' Lady Grace Manners at play – years after her death

Time slips or earth 'memories'?

The cases discussed so far could be interpreted in quite a different way to the time-slip theory. Are percipients slipping back through time to observe events which are *happening*, or observing a visual record which has been accidentally triggered?

Parapsychologists developed the theory of the 'stone tape' to explain a certain category of ghost sightings. This includes cases of ghostly figures, buildings, landscapes and sounds from the past. A good example of the latter occurred near Dieppe in France, on 4 August 1951.

Mrs Dorothy Norton and her sister-in-law, Ms Agnes Norton from England, were on holiday in the village of Puys, near Dieppe. Their second-floor room looked out a short distance from the beach. At 04.20 hours on the above date, Agnes climbed out of bed and left the room presumably to relieve herself. On her return she asked Dorothy, who was also awake, if she could hear a 'noise'? Dorothy said she had been listening to it for about twenty minutes. They both lay in bed listening to the extraordinary sounds apparently emanating from the beach. Dorothy later described them as 'a roar that ebbed and flowed'.

Although there was no doubt where the noise was coming from, when the women stepped out onto their balcony and gazed in the direction of the beach they could see nothing. The noise grew in volume until they were able to distinguish different sounds. They heard gunfire, shell fire, aircraft dive bombing, landing craft, and the cries of men. As the sounds continued over the next three hours both women made notes, speculating that they were experiencing something paranormal. Dorothy was no stranger to psychic experiences. Five days earlier she had been awoken by similar sounds, although they seemed much further away.

After the night of the 4th, the women wondered if they had heard the battle sounds of the Dieppe raid of 19 August 1942. On that date British and Canadian armed forces attacked the German positions at Dieppe, but it was a bloody failure. More than half of the Allied troops were killed, wounded, captured or missing.

The case was investigated by G. W. Lambert

and Kathleen Gay of the Society for Psychical Research (SPR). They found some discrepancies, but by and large the women's detailed notes matched the timing of the raid some nine years previously. Critics pointed out that the ladies possessed a guide book which included an account of the battle, and that a dredger had been operating in Dieppe harbour that night. Puys lay on the route of a regular London-Paris-Turin flight path. Had the women, primed with expectancy, distorted the sound of a dredger and civilian aircraft into the noise of a World War II battle?

The women claimed they had not read the book until curiosity drove them to look at it several hours after the noises had started, and, as investigators discovered, they only had a general knowledge of the affair. A member of the SPR reported that he had camped to the east of Dieppe later that same month in 1951, where he had been awoken by the dredger. He described it as sounding 'like a zoo gone mad'. Would the women have heard the machine some two miles away, and could the infrequent civilian aircraft have accounted for the protracted 'dive bombing' the women had heard? Yet if the sounds were unusual, why was not the whole area alerted?

One is reminded of Joan Forman, who when she moved away from the spot where she had been standing, lost sight of the four children. Did she *have* to be standing in exactly that spot, and under certain other conditions, for the scene to play for her? Perhaps the battle noises could only be heard in the hotel room at Puys – and in the company of Dorothy Norton, who may have been a psychic channel.

The stone-tape theory postulates that certain events, especially those which generate a lot of emotional energy, somehow become imprinted in the environment, such as the stone walls of a building, the ground, or the atmosphere. This could be applied particularly where an apparition has been sighted in the same location over a period of years. For most of the time nothing is there, then certain conditions, such as electromagnetic energies in the atmosphere, or the arrival of someone with developed psychic 'powers', could act as the trigger which sets the recording off.

This seems to negate the idea that the percipient has travelled back in time, but the theory can only be applied to cases where the apparitions seem oblivious to their observers. What of those experiences where there is an interaction between observer and the observed? Then we cannot be dealing with recordings of the past but perhaps the past itself.

Adventures in time

In 1973, a coin collector from Norfolk not only claimed to have slipped back in time but produced physical evidence to back this up. A fellow numismatist told him of a shop in Great Yarmouth which sold small plastic envelopes ideal for holding single coins. He had a general idea of the shop's location and found the likely place in a cobbled street.

Its gaily painted façade made an immediate impression. Inside, the shop was incredibly old fashioned. The young female assistant added to the atmosphere of a former age. She wore a long black skirt and a blouse with leg-of-mutton sleeves, and her hair was piled on top in a bun. Strangely, the shop was silent, with no traffic noise filtering through into the building.

The customer asked for the small envelopes and the girl turned to a box containing a large number of the objects. He remarked on this, considering their scarcity, and she told him that men from the sailing ships bought them for their fish hooks. He was charged a 'shilling' and he handed over a 5p coin, which the assistant looked at with some surprise but made no comment.

He returned a week later for further envelopes but discovered the cobbles had now gone and the decor and interior of the shop changed. A mature female assistant denied any knowledge of the young girl and replied that they did not stock the small plastic envelopes. The manager backed up her statement.

Fortunately the customer still had the original envelopes. They had been placed in a paperbag with the shop's name printed on it, but after a few days the bag had fallen into bits and was burnt. The envelopes also aged quickly, turning brown and becoming brittle. Joan Forman who investigated the case had one of these examined by the manufacturers. They estimated that the condition of the item indicated it was between ten and fifteen years old. It was made from cellulose; the firm had first developed the cellulose process during World War I but it was during the 1920s that the envelopes went into production. The man thought the girl was wearing the fashions of the

Edwardian era, and put a date of between 1900–1912 on the incident, which lends confusion when compared with the date of manufacture of the envelopes.

The Versailles affair

The most famous case of time dislocation occurred to two English tourists visiting the Palace of Versailles, home of the French royal family during the seventeenth and eighteenth centuries. The adventurers, as in the Dieppe case some fifty years later, were two women: a Miss Anne Moberley and Miss Eleanor Jourdain. These middle-aged ladies were academics of some standing. Miss Moberley was the Principal of St Hugh's College, Oxford, and Miss Jourdain head of a girls' school in Watford. Both were interested in history and not prone to fantasize.

On the warm afternoon of 10 August 1901 after leaving the Galeries des Glaces the spinsters decided to walk through the grounds to the Petit Trianon. Not sure of the way, they took a side lane where Miss Moberley saw a woman shaking a cloth from the window of a building. She learned later that her friend had not seen this, indeed the building did not exist. On a path they came across two men wearing long greyish-green clothes and three-cornered hats. They seemed to be working on the path because a wheelbarrow and spade were nearby. The men gave them directions and the ladies continued their walk.

Then Miss Jourdain noticed a woman and a teenage girl standing in the doorway of a cottage, dressed in old-fashioned clothing. At this point the landscape took on nightmare proportions, becoming flat, almost two dimensional, and both women felt a wave of depression sweep over them. Coincident with this they approached a circular garden kiosk where a man was sitting. He looked sinister and repulsive, and they would not pass him by. Suddenly there was the sound of footsteps behind them, but when they turned, no one was there. Miss Moberley noticed someone else standing nearby, a warm smiling man wearing a cloak and sombrero. He directed them to the house.

On the way Miss Moberley noticed a woman sketching on the lawn. She was wearing a dress with a low-cut neckline and a white shady hat. The woman turned and looked at the women as they walked past. Only afterwards did Miss Moberley discover that her friend had not seen the figure who bore a striking resemblance to the eighteenth-century Queen of France – Marie-Antoinette.

As they turned the women noticed a young man 'with the air of a footman' emerging from a building. He slammed shut the door, and directed them to the entrance of the Petit Trianon. Once inside, the depression and atmosphere of unreality passed. Had the women slipped back in time and observed buildings and people dating just before the French Revolution, or is there a more prosaic explanation? Their book, *An Adventure* was published ten years later. Since then the case has been exhaustively investigated.

Critics found discrepancies in the women's accounts. Further, it was discovered that an aristocrat called Comte Robert de Montesquiou-Fezenzac, obsessed with the eighteenth century, used to dress up in that period and with some friends frequent the gardens of Versailles. Someone also claimed that as a child she knew of a woman who in the summer often dressed up as Marie-Antoinette and sit in the garden of the Petit Trianon. Had the two English ladies merely come across actors in period costume? If one accepts this as the explanation the other intriguing facets of the story are ignored. If there were actors there, how was it on several occasions that only one or other of the women saw them?

The ladies described buildings and followed paths which no longer exist in the twentieth century. Indeed, to have followed the route they claim to have taken entailed walking through several brick walls. The pall of depression, the sensory effects, have all been described by other experiments of the time dislocation phenomenon.

Room at the inn

Another apparent time slip, again involving English tourists in France, took place in October 1979. Len and Cynthia Gisby and their friends Geoff and Pauline Simpson intended travelling to Spain from their home in Kent. After crossing the Channel they took the autoroute to Montélimar. As darkness began to fall they stopped at a hotel called the Ibis, but in the reception hall a man wearing a plum-coloured uniform told them there were no rooms available, but if they took a back road they would arrive at a small hotel which he was sure would have accommodation.

While walking through the grounds of the Palace of Versailles towards the Petit Trianon (pictured above), two English ladies apparently slipped back in time

They found the road but before long it had badly deteriorated. The women noticed some posters advertising a circus, which were very old-fashioned in style. Eventually they found the hotel, and pulled in at the side of the road because there was no car-park. Alongside it was another building which looked like a police station. Although the hotel owner spoke no English, and they spoke little French, they were able to make themselves understood and found that there were rooms. It was now 22.00 hours.

The ranch-style two-storey building was very old-fashioned inside. Their bedroom windows were unglazed and shuttered, bed sheets were of heavy calico with bolsters instead of pillows, and the bathroom had plumbing more suited to the Victorian era, with the soap stuck through with a spike.

After unpacking they went downstairs and ate a large dinner heaped on metal plates, washed down with beer. Several men in rough clothing sat at the bar. In the morning after a good night's sleep the four came down for breakfast. While they were eating, a woman walked in with a small dog under her arm. She was wearing a long ballgown and buttoned boots. Then two gendarmes entered dressed in tall peaked caps, deep blue capes and gaiters.

By now the Gisbys and the Simpsons were convinced that they had stayed the night in a working museum – a creation to entertain tourists. They decided to take some photographs. The men each photographed their wives leaning out of the bedroom windows. Now it was time to leave. First Len and then Geoff tried to get some instructions off the gendarmes as to the direction of the autoroute, but no matter how they pronounced it the officers seemed not to understand the word. However, when they mentioned 'Spain', they were directed towards the old Avignon road.

There was more confusion when they went to pay their bill. It was less than the equivalent of £2! Len's protestations only drew smiles from the patron and the gendarmes. Eventually they left, and instead of using the Avignon road, studied a map and found the autoroute with no trouble. They headed for Spain, and two weeks later were coming back. It made sense that they should choose to stay another night in the quaint and cheap hotel near Montélimar.

They found the road, and even saw the circus posters, but there was no hotel. An exploration of the area drew a complete blank. Bemused, they drove to the Ibis and asked to speak to the man in the plum-coloured uniform. They were informed that no one of that description worked for the Ibis. Enquiries of the location of the hotel itself drew a blank from the staff.

Back in England the holiday snaps were processed. The friends were surprised when the three photographs taken at the hotel were not included. Surprise turned to incredulity when an examination of the consecutively numbered negatives showed that no exposures had ever existed. One camera showed some mechanical evidence of having tried to wind on but failed. However, exposures were missing from both films in two separate cameras.

In 1983 the couples returned to France to make a thorough search with the help of the French Tourist Board. The tourist board's representative, Philippe Despeysses, had found a place which corresponded in some respects to the mystery hotel. The Gisbys and Simpsons were taken there, and although they admitted that it was very similar, a chat with the owners convinced them it was not the place they had stayed at in 1979.

Jenny Randles interviewed the couples about their experience. She found other problems with the story apart from the missing exposures. If they really did slip back into the past, she asked, how was it that no one in the hotel found their car or clothing unusual? Why did the patron accept payment in coins which would obviously be worthless in a former time? The Simpsons, who seemed sincere and convinced, replied, 'You tell us what the answer is. We only know what happened.'

The land behind the trees

Elinor de Torri Hudson lives in Florida, USA, but in June 1939 she was touring with an opera company around the small, often poor, towns of South America. It was on a hot sunny day in that pre-war year when she found herself with free time. As a result she entered a nightmare.

Along with Inez Bertolli, the prima ballerina with the touring group, Elinor had decided to explore their current stop, Santa Maria in Argentina. The company was scheduled to put on a performance that day but neither were involved in that particular production. So they started to wander the small streets and a park in the centre of the big village. After idly watching the children play on swings they decided to explore the far side of this open space. They had to push through trees and shrubbery but these moments of effort were rewarded by an unexpected sight. For there was nothing except empty land overgrown with weed, stretching ahead in front of them. It seemed completely out of place. What was more the two women noticed an eerie stillness and silence descend over them. All sounds, of children laughing, birds singing, indeed everything, had just ceased. It was as if they were on another world.

Inez grabbed Elinor's arm saying that she felt awful. Even as they spoke a noise intruded. It sounded like a cart rolling towards them across the empty pampas. They heard it clearly – including the snorting beast that pulled it, the creaking wood and rattling wheels. It passed almost right through them as they stood rooted to the spot. But nothing was visible.

As they flung themselves aside from a collision with the unseen terror a man cried, loudly and clearly from only feet away, '*Vamos*!' Then he cracked a non-existent whip and the noise completely disappeared.

Inez, full of South America superstition, was convinced the sound was an omen of death. She wanted to flee the town. But Elinor persuaded her to stay and nothing untoward happened. The tour proved a great success.

However, back at their hotel they mentioned the strange land behind the park. The manager stared at them and nodded. Nobody went there now. He explained that centuries before a village stood there but was ravaged by plague. It was totally wiped out and then finally razed to the ground by fire. The many dead were taken away in heavily laden ox-led carts.

To the future

Time slips into the future generally lack the physical involvement of the percipients demonstrated in some experiences of the past. Future events often make themselves known through precognitive dreams, or during other states of altered consciousness entered while involved in carrying out a repetitive task, or in an extremely relaxed state. One case however, where a participant did visit a future landscape, was recorded by Joan Forman.

A university professor told her of his experience in 1896 when he was seven years old. That morning he had been flogged at school, so in the afternoon he decided to play truant in the streets of Hanley, Staffordshire. He remembers vividly scraping his iron-shod boots on the pavement, generating sparks, and, oddly, stopping to listen to organ music emanating from a church which was empty and locked up.

He went down a back street which was familiar to him. It ended with a blank wall beyond which lay some waste land. But not on this particular day. There was a doorway in the wall. The boy went through and found himself in a different world.

He was in a small town consisting of buildings he did not recognize. Walking along one street he entered several houses but they were all empty. Eventually he climbed upstairs in one and looked out across a valley that terminated in tree-covered hills.

What happened after that is not clear. Twenty-one years later he was in the armed forces serving in France, near the Somme, at a time when the Germans were in retreat. One evening he had some free time and decided to visit one of the local villages and do some sketching. He came to a village called Malaunay (Misery) and immediately recognized it from his boyhood experience.

All the inhabitants had fled. He entered several of the houses but left quickly when he discovered the bodies of dead German soldiers. Then he found a house empty of such horrors and went upstairs. The view was exactly as in his boyhood experience.

Had the professor slipped forward in time twenty-one years in 1896? If so, where were the German bodies? Another discrepancy involves the name of the street in the original 'vision'. It was called Windmill Street – a very English name. When the professor returned to Hanley some years later, he discovered that some houses now occupied the waste land. Some of them were on a road called Windmill Street.

The professor seemed to have been open to two future reference points that morning in 1896. One concerned his own personal future, and the other the future of the waste land in the back streets of Hanley.

Time television

In 1974 a writer named Andrew Tomas published a book called *Beyond The Time Barrier*. In it he talked about the possibility of 'time television' – when electromagnetic 'pictures' from the future might manifest themselves on television sets. In the same year of publication an incident occurred, apparently demonstrating the reality of time television, too late for inclusion in the book.

On Saturday, 1 June 1974, the giant chemical plant at Flixborough now in Humberside, exploded in a devastating ball of flame. The accident destroyed most of the plant and killed twenty-nine people. The surrounding area was evacuated as huge black clouds containing poisonous gases were released by the terrific heat. Yet one person, housewife Lesley Castleton, heard of the tragedy six hours before it happened.

'I was watching the Saturday morning film on television,' she later told interviewers, 'when it was suddenly interrupted by the words "News Flash" which came across the screen. A man's voice gave details of a horrific explosion at Flixborough. He gave numbers for the people who had died and been injured, and the

The giant chemical plant at Flixborough exploded on Saturday 1 June 1974 at 4.30pm, but Lesley Castleton 'saw' a news flash describing the incident six hours earlier

names of chemicals responsible for the explosion. The News Flash then went off and the film continued.'

Mrs Castleton was utterly convinced by the news broadcast, and when some friends came to see her at lunchtime she told them of the awful tragedy. When she and her husband watched the news that night, it stated the time of the accident as 16.30 hours. Mrs Castleton naturally assumed that the reporters had made a mistake. When she read the Sunday morning paper the following day, and they confirmed the time, she realized something was very wrong. Mrs Castleton telephoned her friends and they assured her she had told them of the accident that lunchtime. 'I went really cold. I realized then I had seen something which hadn't yet taken place.'

Was this a television broadcast from the future slipping through a crack in time? If so, why that particular item, and why did it fail to appear on other television sets in the region? Could it have been a premonition which presented itself in the form of a television broadcast to make it acceptable to Mrs Castleton?

Early in 1946 Helen York of Bellwood, Illinois, was washing up the dinner pots when the glass in the kitchen window shimmered, and the scene outside subtly changed. She told *FATE* magazine in January, 1988, 'There on the glass, like a movie, I saw the image of our Buick which was parked out in front of the house. Then, in the picture, I saw a car, speeding from the north, sideswipe our car. The driver pulled over to the curb and a man staggered out, looked at our smashed car and started to run.'

But that was not the end of it. The 'camera', like a movie, followed the man into a house where he hid in a basement behind some boxes. The viewpoint changed and Mrs York saw the police examining their car and the man's vehicle and then radioing through the information. Next, the scene cut to the front of the police station where officers were taking the driver inside. There the desk sergeant booked him with drunken driving and leaving the scene of an accident. The glass shimmered again and the scene outside returned to normal.

Mrs York said nothing, and she, her husband, and a friend called Smitty, went out to attend a political meeting. They went in the friend's car, and despite her protestations, the couple's Buick was left parked on the road.

They arrived home late to discover that someone had run into the parked car. The men suggested calling the police but Mrs York surprised them by stating that they already knew about it, and had the man in custody. When they spoke to the desk sergeant, he confirmed the details exactly as Helen York had observed them several hours before.

A timetable of the facts and conclusions

- Our physical selves operate in the three dimensions of space only, but consciousness moves forwards and backwards through time.
- Time-slip percipients often report the sensation of existing in two time zones simultaneously – one overlapping the other.
- The absence of natural sounds, such as bird song and traffic noises at the onset of a time slip, have also been reported by percipients of other phenomena such as UFO close encounters.
- Time slips seem to require a 'trigger'. This could be a sudden bright light, or abnormal amounts of electrical energy in the atmosphere which, if the conditions are exactly right, interact with the brain of a percipient.
- Time slips are not imaginary constructs. Often the information is discovered to be entirely accurate.
- The branch of physics known as quantum mechanics may hold the answer to the true nature of time.
- J. B. Priestley in his book *Man and Time* divided time into three components: *Time One* clock time, *Time Two* the time of the possible future, *Time Three* the time of the imagination.

DEATH BY SUPERNATURAL CAUSES?

SPONTANEOUS HUMAN COMBUSTION

The phenomenon of spontaneous human combustion is arguably the most neglected scientific mystery of our times. Some professional fire fighters and scene-of-crime police officers assert the phenomenon exists and this should be enough to instigate a thorough scientific study. If it is true that the human body can erupt into flame for no rational reason and in most instances incinerate to ash, the process needs to be understood so that we can prevent it occurring and if a natural energy is involved, harness its potential for other purposes. In spite of this lack of research the sceptics smugly assert that existing fire science can explain away every case of alleged spontaneous human combustion (SHC).

At the beginning of 1985 Peter Hough and Jenny Randles had only a peripheral interest in SHC. Then a teenage girl burst into flames at a Cheshire college, and the media dubbed it 'spontaneous human combustion' after a comment made at the initial rapidly convened February inquest. They began making their own enquiries and were subsequently invited by the police to the new inquest held in June. What they witnessed was a wealth of conflicting evidence, a poor official investigation and a one-dimensional interpretation of the facts. A Home Office chemist reported that the girl had caught fire by leaning against a lighted gas ring during a cookery examination. Her smock had smouldered, then burst into flames possibly ten minutes later when she was in another part of the building on some stairs and between floors.

On the face of it this sounded an attractive explanation. Indeed, many of the journalists present left early and wrote up their stories with this conclusion. However, it ignored several facts. The gas ring which was lit was a back ring, none of the twenty or so witnesses saw smouldering or smelled burning, and the girl had felt no discomfort up to the moment when her back erupted into flame 'like a stunt man on TV', as one witness described it. Nevertheless, the jury, kept ignorant of any other possible explanations, brought about a verdict of 'misadventure'.

Only afterwards was Peter Hough contacted by a senior fire officer who informed him that the coroner had refused as evidence a thirty-page report compiled by the Cheshire Fire Brigade and the prestigious Shirley Institute in Manchester. While not endorsing SHC the report refuted the cooker theory. As the authors said in their first book *Death By Supernatural Causes?*, 'In truth, on the evidence which has been made public, there is no basis at all on which to state unequivocally what caused a pretty teenage girl to burst into flames, surrounded by fellow students. An 'open verdict' would have been a more fair and accurate reflection of the evidence.'

The authors' own researches had to start virtually from scratch because of the lack of recorded data on the subject. They contacted forensic scientists, police and fire officers, followed up new cases, pursued fresh information on old ones and talked to people who had survived SHC. Along the way they found coroners who were friendly and cooperative but also coroners who were deliberately obstructive. Sometimes the most innocent of facts were buried in a web of secrecy. Bureaucrats denied them information they could have freely obtained had they been present at inquests.

A history of phenomenal burnings

The first recorded case alleging SHC occured on 26 June 1613, in Christchurch, Dorset. Carpenter John Hitchen went to bed with his wife and child after a hard day's work. Mrs Hitchen's mother, Agnes Russell, sleeping in a separate bed, was awakened by a terrible blow to her cheek. There was an electrical storm going on at the time. She cried out and went over to the other bed and awoke her daughter who was burnt all down one side. Mr Hitchen and the child were dead beside her. Mrs Russell and her injured daughter dragged him out but into the street, where he continued to burn for three days, reducing to ashes.

During the nineteenth century SHC was a subject for hot debate. One proponent was Charles Dickens who in *Bleak House* destroyed his character Krook through spontaneous combustion. Dickens studied thirty cases of the phenomenon to make the fictional death as accurate as possible. He was attacked by sceptics who accused him of perpetuating a myth. The explanation promoted by the sceptics was in

When Charles Dickens described the spontaneous human combustion of Krook in *Bleak House*, critics lambasted him for it. It is still a subject which arouses great controversy

time to become a myth itself. They stated that all the victims of mysterious conflagration were 'drunkards', and that alcohol absorbed into the skin rendered it exceptionally combustible. An accidental fall into an open grate, or brushing an arm against a candle flame could set it instantly ablaze, reducing the victim to ash. Subsequent experimentation on corpses by Baron J. von Liebig, famous for isolating titanium, proved that alcohol could not be absorbed and retained in the fabric of the body.

The authors' own researches show that the 'drunkard' appellation was probably just a convenience adopted because it fitted in with the sceptical theory of the time. It seemed that any victim who was known to enjoy a drink, was instantly deemed an alcoholic. The case of Grace Pett illustrates this point perfectly.

Grace Pett was the pipe-smoking wife of an Ipswich fisherman, who at sixty, met her death on the night of 9 April, 1744. An idiosyncrasy of the lady was to come downstairs during the night and sit by the fire smoking her pipe. On that particular night, her daughter, who slept in the same bed with her, did not perceive her rise and was not aware she had gone until the following morning. After dressing, the daughter went downstairs and found the remains of her mother in the kitchen.

The woman was stretched out on her right side, head nearest the grate, body extended across the hearth, legs across the wooden floor beyond. Apparently the remains had the appearance of a wooden log, baked rather than ravaged with flame, still glowing. The trunk was incinerated, resembling a heap of coals covered in white ash, and the head and limbs were also burned.

There was no fire in the grate, and a candle, close by, had burned down entirely in its candlestick. Although fat from the body had so penetrated the hearth, 'it could not be scoured out', the floor was not even discoloured. Near to the consumed body were some children's clothes and a paper screen. These had suffered no injury. It is interesting, considering that smoke inhalation is the prime cause of death in fires, that the daughter had not been aware of a conflagration until she had sight of her mother. There are indications too that there were other people staying in the house that night, none of whom had been aware of a fire.

Once again, in later reports, much was made over the 'fact' that Mrs Pett had consumed 'a large quantity of spirituous liquor' that night to celebrate the return of another daughter from Gibraltar. However, through the diligent researches of historian Peter Christie, we learned the truth.

After much effort Peter received from the Royal Society of Great Britain a copy of a letter dated 28 June 1744 written by a Mr R. Love – apparently the earliest documentation on the case. Details of this letter are given in a journal called *Philosophical Transactions* which the present authors had the good fortune to track down.

Mr Love attended the inquest and learnt that the women retired at about 10 p.m. and the body was found at 6 a.m. The letter stated there was no fire in the grate, and contrary to some subsequent reports, Mrs Pett 'was not in liquor nor addicted to drink Gin'. It confirmed that the extremities and parts of the head were unburned, and that the wooden floor was not even scorched. The bones were calcined so completely that the remains were easily shovelled into a coffin. Despite all of these anomalies, the jury brought about a verdict of 'accidental death'.

Grace Pett is a good example of a victim reported at the inquest as not addicted to alcohol, who was not drinking on the night of her death, but in later legend was conveniently said to have consumed 'a large quantity of spirituous liquor'.

A fire officer's investigations

During our enquiries we came into contact with several senior fire officers. One was Tony McMunn. Tony has been a serving fireman for twenty-five years He spent time as Assistant Divisional Officer, lecturing to colleagues at the Fire Service College in Gloucestershire, about Fire Prevention, and unofficially on spontaneous human combustion. Currently he is Deputy Chief Fire Officer for Hereford and Worcester. Tony's inauguration into the SHC controversy was a true baptism by fire. At the time he was attached to the Lancashire Fire Brigade.

'Around 12.30 midday, on 4 March 1980, I received a call to go to a "persons reported" incident. That is Fire Brigade jargon for someone trapped in a house fire, or a fire fatality.' The address was a terraced house in Chorley, inhabited by an elderly lady. When Tony arrived, he expected to see lots of smoke and flames but strangely there was no sign of the fire he had come to see. A police officer who met him there, said, 'You've got to see this to believe it!' Tony entered by the front door and walked into a back sitting room which contained a bed. What he saw on the floor left him aghast. It was the remains of the old lady.

'The woman was found with part of her body in the fireplace and the lower part on the carpet. The knees to the skull were just calcined remains – in other words, pure ash. From the knees to the feet the legs were intact. There was no damage by fire to the surrounding area, although there was smoke damage from the upper part of the walls to the ceiling.'

The approximate time of death was deduced by the fact that passers-by and neighbours had noticed a large volume of smoke and sparks issuing from the chimney at approximately 21.30 hours the previous evening. However, it was almost fifteen hours later before a neighbour called round to see the pensioner and discovered the bizarre tragedy.

There was a bucket in the room. We presumed that while she was relieving herself, the victim had collapsed into the hearth where there was an open coal fire. When I first saw the phenomenon, I was taken aback. I had never seen anything like it before, neither had the ambulancemen nor the police, although like all other Fire Brigade personnel, I had attended many fatalities involving fires. Even the bones were gone, yet when you look at nearby objects such as a brushed-nylon footstool and clothes to the right of the remains, these were just stained.

At the time I hadn't a clue what it was. Someone mentioned to me it was 'spontaneous human combustion', and I naively accepted the explanation without knowing exactly what the terminology meant. It was later that I began to wonder what the phenomenon might be.

The inquest was held at Chorley Magistrates Court where a verdict of 'accidental death' was recorded.

Tony became intrigued and spent the next decade researching the phenomenon and gathering material from here and abroad. This was done in his own time and at his own expense. Some material came from colleagues who had similarly attended incidents where victims had been reduced to ashes. The temperatures necessary to bring this about should turn the immediate environment into an inferno, yet in case after case, combustible items only inches away remain just scorched.

Q.E.D.

What was it that could reduce even bones to ashes yet leave surroundings relatively untouched by fire? Some scientists and fire experts believe they have the answers in existing fire science. The theories were exploited by the BBC documentary series *Q. E. D.* , in a programme entitled 'A Case of Spontaneous Human Combustion', broadcast on 22 April 1989, and subsequently shown in the USA. They revolved around what is termed the 'wick' or 'candle effect' and, incongruously, oxygen starvation in a sealed room. David Halliday, a forensic science expert for the Metropolitan Police Force, stated that in every case he had examined there was always a good ignition source nearby.

There are possible ignition sources in almost every room of a house. But does that mean they were the cause, or indeed that in the case of fires and cookers, they were alight at the time? David Halliday reasons that victims become drunk, or suffer fatal heart attacks and then fall against lighted appliances. Once lit, the candle effect takes over. Dr Dougal Drysdale based at Edinburgh University explained in the programme, and later in an interview with the authors, how this works.

'In a way, a body is like a candle – inside out. With a candle the wick is on the inside, and the fat on the outside. As the wick burns the candle becomes molten and the liquid is drawn onto the wick and burns. With a body, which consists of a large amount of fat, the fat melts and is drawn onto the clothing which acts as a wick, and then continues to burn.'

In the programme Dr Drysdale demonstrated the theory using a sausage of animal fat wrapped in some cloth. It was lit at one end and using time-lapse filming burnt through to the other end. But what about the bones? In most SHC cases even bones are reduced to ash.

The wick effect experiment had already been demonstrated on television in an item on SHC in 'Newsnight', on BBC 2, in January 1986. Then it had been carried out by Professor Gee, Emeritus Professor of Forensic Medicine at Leeds University. Bones were not included then, either, although the professor glibly assured the interviewer that if there had been, he was sure they would have reduced to ash – a claim that no one has been able to demonstrate. What did Dr Drysdale have to tell us? Did he believe that his experiments had disproved SHC, as the *Q.E.D.* programme hinted?

'I have no experimental data to prove that it can or can't happen. There are overwhelming scientific arguments against it, and there is no mechanism by which it could occur. Personally, I believe that the "wick effect" can account for the total body destruction. Ignition is not spontaneous but is external.'

Nevertheless, Dr Drysdale confirmed that while the theory has been tested in the laboratory, 'no one has been able to convert the bones to ash under these conditions.'

Did this not argue strongly against the wick theory as an explanation for SHC? The reduction of bones to ash is perhaps the most difficult factor to explain. Apparently, even the demonstration did not go as smoothly as the programme-makers led the viewers to believe, as Dr Drysdale confirmed.

The original piece of fat supplied by the BBC would not burn, and another piece was tried. What the viewers saw was an edited piece of filming while Dr Drysdale endeavoured to get the experiment to work. He told us, 'It took a long time in this instance. It may well take a long time in such tests, but in the end it will go. Of course, it depends upon the amount of fat in the sample. The type of cloth used also seems to be important. Some types are better than others.'

Did he think this meant that corpulent people would be more likely victims of the wick effect?

'That is the implication – yes. And many are. But not all victims are fat. There are instances where the victims are lean.'

But even if the wick effect could be made to work, not only on a piece of pure fat but on an organism composed of between 60–70 per cent water, how could the fire be contained in the immediate area, without spreading to surrounding furnishings? Stan Ames, Head of Reaction to Fire Section, of the Fire Research Station in Borehamwood, thought he had the answer.

In the programme Mr Ames postulated a scenario of a room with its doors and windows shut. A person in the room collapses against a source of ignition and their clothes catch alight instigating the wick effect. Since the room is sealed, the oxygen level drops to around 16 per cent. At this level, the body continues to smoulder but insufficient oxygen will stop the fire spreading.

In order to test this hypothesis, an armchair, substituted for a body, was placed inside an air-tight chamber with other objects, and then set alight and left for six hours. The narrator informed the viewers that the demonstration would prove that a body could be reduced to ashes, 'or in this case an armchair to its springs'. What was promised did not materialize. The seat and back of the chair, although badly burned were still remarkably intact. Certainly it was not reduced 'to its springs'. In any case, how can an object composed of highly inflammable material be substituted for organic tissue? We learned that it had taken all morning to get the chair to burn at all.

The 'wick effect' and oxygen depletion scenario has not yet been demonstrated as an effective explanation for alleged SHC deaths, and remains nothing more than a theory. There is doubt too in the premise that victims simply drop down dead near a source of ignition, whereupon the above hypothesis takes over. Post mortems on many victims show that they died of asphyxia due to the inhalation of fire fumes. These people were alive when they began incinerating, which begs the question why could they not put out the fire? In any case, the sceptic's 'explanation' only applies to cases of people found incinerated in an air-tight room.

What the cremation experts say

Cremation experts have been scathing about the explanation for SHC promoted in the *Q.E.D.* programme. Mrs Valerie Bennett is a local government officer and a qualified superintendent of Overdale Crematorium in Bolton, Lancashire. Hough showed her and her staff several photographs depicting SHC victims. They were shocked by the pictures. Not emotionally, but professionally. They could not believe the utter devastation and the lack of any skeletal remains. Mrs Bennett shook her head as she handled the photographs. 'We are cremating bodies between temperatures of 600°–950°C for on average one and a half hours, and bones, not ash, always remain.'

But what if the cremation process was continued for several hours, would that reduce the bones to powder?

'All that does is burn the bones further and turn them black. You would still end up with bones – not dust. The larger bones are still recognizable – such as the pelvis and thigh bones.'

There is a final process which does reduce the bones to dust. The remains are put into a 'cremulator' where they are pounded into dust by eight heavy iron balls.

What did Mrs Bennett think of the oxygen-starvation hypothesis to account for the other unburned items in the room. It was claimed that the body would slowly incinerate under these circumstances. Yet the cremation process occurs under exactly opposite conditions. During cremation, the cremator is continuously fed copious quantities of fresh air under pressure by the use of fans. If this was not done, then what would be the result?

'The body would be hardly burned,' she replied. 'I just cannot see how a human body could generate sufficient heat to turn a room into a cremator.'

Head cremator, Peter Thornley added, 'If we turned the oxygen off, the body would just go black. It wouldn't continue to burn.'

Mrs Bennett concluded, 'I cannot see that a fire in an ordinary room can achieve what we cannot under intense heat and controlled conditions. We've had cases where someone has fallen on a fire and died. Their clothes have

After cremation, the bones are put inside a cremulator (*above*) with eight iron balls weighing several pounds each. A drum revolves, and the bones are pounded into ash. Cremation experts scoff at the idea that SHC incidents can be explained in mundane terms. Theories put forward by the BBC *Q.E.D.* science programme did not stand up when investigated by the authors

Head cremator, Peter Thornley (*left*), operating one of the five cremators at Overdale in Bolton, Lancashire. Temperatures up to 950 degrees centigrade are required here to incinerate a human body. Even so, bones still remain

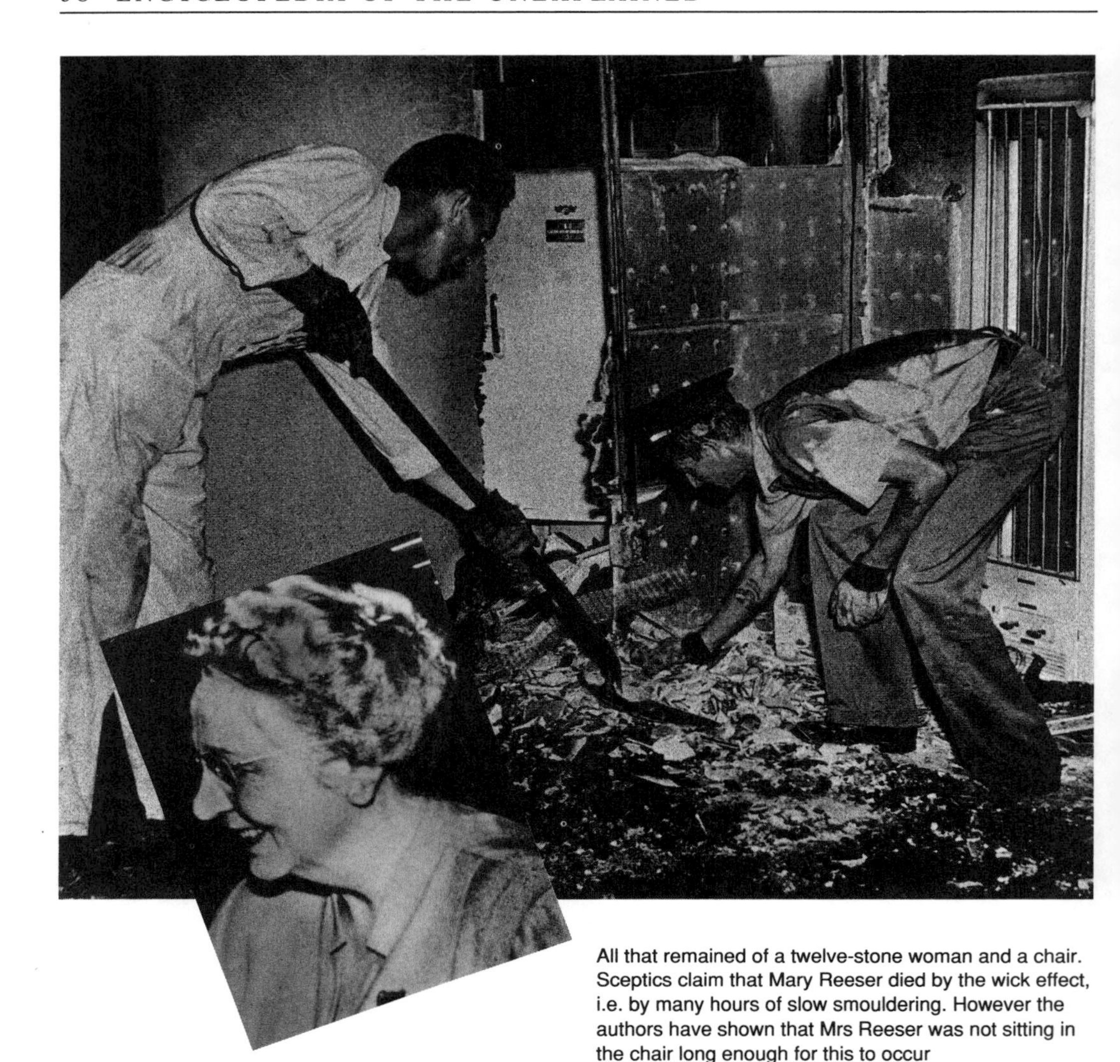

All that remained of a twelve-stone woman and a chair. Sceptics claim that Mary Reeser died by the wick effect, i.e. by many hours of slow smouldering. However the authors have shown that Mrs Reeser was not sitting in the chair long enough for this to occur

caught alight, but no one has ever burned right through. There is always flesh and a skeleton left.'

Peter added, 'Even in the cremation process, the skull is incredibly hard to destroy, yet in many of these [SHC] incidents, even the skull is gone.'

The classic case of Mrs Reeser

What of the classic American case of Mary Reeser? Both Mrs Reeser and the armchair she was seated in were found almost completely incinerated to ashes, including the bones, on Monday, 2 July 1951. The fire did not spread to the rest of her apartment, despite the fact that, according to a telegram boy, the windows were open. The authors' investigations, which included the study of interviews carried out with witnesses by FBI agents, dealt a blow to the sceptics' supposition of the wick effect which would account for Mary Reeser's death. The wick effect theory requires many hours of slow smouldering to work. Sceptics, such as American researchers Joe Nickell and John Fischer, have often stated that Mrs Reeser fell asleep in the armchair smoking a cigarette, thus giving the wick effect twelve hours to work. A careful study of police records shows that Mrs Reeser did, in fact, retire to bed, then got up in the early hours possibly to get some fresh air. She then sat in the armchair not later than 04.20 hours where the conflagration overtook her – the time when fire stopped her electric clock, just a few hours before her ashes were discovered.

SHC out of doors

There have been many recorded instances of people bursting into flame out in the open air. Dr George Egely of the Physics Institute in Budapest, Hungary, reported a tragic case which occurred on 25 May 1989. An engineer and his wife were motoring near the village of Kerecseud when the man decided he had to urinate. He stopped the car and walked over to some bushes. Suddenly, his wife was faced by an horrific sight. Her twenty-seven-year-old husband had become surrounded by a blue aura. He threw his arms out wide and collapsed onto the ground.

As the distressed wife attended her dead husband a bus containing medical men travelling from a convention stopped. One of the victim's trainers had been stripped off by a burst of energy, burning a hole in it. An autopsy later revealed that his internal organs in the abdomen had been carbonized.

Computer operator Paul Castle was luckier. He was one of several victims of SHC who survived to tell the tale. The teenager was walking home in 1985 through Stepney Green, London – coincidentally also on 25 May – when fire suddenly engulfed the upper part of his body.

He described how it was like a huge gas jet suddenly being lit, as if he had been doused with petrol and set alight. The torture of the flames drove him to run in panic, then he fell to the ground and curled up into a ball. The flames stopped as suddenly as they had appeared. It was as if they had been turned off. Paul staggered into nearby London Hospital where he stayed overnight to receive treatment for his injuries. They were described as extensive but superficial. Paul Castle was a very lucky young man.

Combustible theories

In their book *Spontaneous Human Combustion* Peter Hough and Jenny Randles documented cases which have occurred in the open air and they record interviews with people who have survived SHC. These people know there was no source of ignition nearby. What then are the possible explanations? These range from little-understood areas of physics to the esoteric.

Ball lightning was a phenomenon unrecognized by the scientific establishment until around twenty-five years ago, and even now many physicists have trouble coming to terms with it. This is because of its contrary nature. These wandering balls of electrically charged plasma give the appearance sometimes of being intelligently controlled. Ranging from a few inches to several feet diameter, ball lightning does not need a thunderstorm to manifest. This rare phenomenon has been observed out in the open, inside buildings and even in the fuselages of aircraft. Incongruously there are instances where the phenomenon has failed to generate heat, yet there are examples where water and glass have been vaporized and metal warped. Could ball lightning form inside the human body and suddenly discharge all its energy at once causing rapid incineration?

A nuclear reaction within the body is another theory. American investigator, Larry Arnold, postulates a sub-atomic particle he calls a 'pyrotron'. More interesting is the controversy surrounding 'cold fusion'.

When Dr Martin Fleischmann and Professor B. Stanley Pons publicly demonstrated the heat energy effects of cold fusion on water in March 1989, it caused great excitement. Here was a cheap and safer way of producing nuclear energy. When scientists failed to replicate the experiments the process was discredited. Now attitudes are very entrenched between physicists on cold fusion, although most Japanese scientists accept cold fusion as fact. Fleischmann and Pons are continuing their research in laboratories owned by the Toyota Motor Company.

If cold fusion does turn out to be a reality, what could this mean for spontaneous human combustion? One physicist, who refused to be named, speculated that potassium in the body was most likely to be involved in a fusion reaction. This substance is concentrated in the brain, spinal cord and skeletal muscles. The process would be very hot, perhaps incandescent, and would be over in seconds or a few minutes at the most. A lack of potassium in the hands and feet would leave them least affected, and the speed of the conflagration would hardly affect surrounding materials. The 'trigger' could be found in strong electrical and magnetic fields.

John Heymer, a former scene-of-crime officer with Gwent police in Wales faced SHC on a cold winter's afternoon in 1980, and has pursued the subject ever since. He proposes electrolysis of water within the body. An electrical charge could break the molecules down into their

Fire fact file

■ It is very hard to discover cases of SHC prior to the seventeenth century. Did a change of diet or something new to the environment come about then, or is it simply a lack of documentation in times when communication was poor?

■ The authors have now documented one hundred and twenty cases, and believe these are only a few of many, many more instances. It is much easier for coroners' courts to grasp at a rational solution to strange fire deaths, than to encourage the belief in SHC. Even so, the authors believe the phenomenon is relatively rare.

■ Statistical analysis carried out on one hundred and eleven cases (admittedly a small sample) showed that slightly more women than men fell victim, that only 3 per cent were under five years old, and that just over half the cases occurred in the early hours (00.00–06.00 hours) of the morning.

■ Leading authorities on SHC are American *Larry Arnold* who has studied the phenomenon for almost twenty-five years, and his sceptical countrymen, *Joe Nickel* and *John Fischer*. In Britain *Michael Harrison* produced the first book on the subject in 1976 which gathered together many historical cases. But it was not until 1992, that the first investigation into SHC was published. *Spontaneous Human Combustion*, by Peter Hough and Jenny Randles was originally published in hardback by Robert Hale, but released in paperback during June 1993 by Bantam.

constituent gases – hydrogen and oxygen. A build-up of static electricity could provide the spark which would start off the conflagration.

Among the more 'supernatural' explanations spontaneous fires associated with outbreaks of poltergeist activity have been noted. The depressed state of mind of some victims has also been commented on. It has been speculated that a 'death wish' in potential suicides causes them to erupt into flame, in an extreme manifestation of 'mind over matter'. We know that a negative state of mind can make us physically ill, can it also cause total destruction? Practitioners of Kundalini Yoga remark on the terrific heat they have experienced during arousal.

There are many more theories, some connected with chemical reactions within the body, which the authors' explore thoroughly in their book. The body contains many volatile substances, including methane gas. Whether you believe in SHC or not, it is undeniable that there are many fundamental questions which still cry out for an answer. The subject needs some serious research. Tony McMunn agrees.

> *There is a phenomenon of which no one has been able to explain. The only way to attempt an explanation would be for someone to be present when it happens. The nearest I've been to that was when I was confronted with the calcined remains of a human being. I can't give you an answer as to why most fire victims just suffer surface charring, and why a very small number are almost totally incinerated. People may say it's because there is a slow smouldering fire, but the truth is that no one really knows. There is something present in these cases that is out of the ordinary and unexplained.*

THE ANIMAL MUTILATION MYSTERY

Strange animal mutilations have been recorded over many decades. They have been blamed on natural predators, Satanists, sadists, poltergeist manifestations, secret government departments and UFO occupants.

Charles Fort, who recorded thousands of anomalous events from around the world at the beginning of the twentieth century, brought to the attention of phenomenalists a fascinating case from Binbrook in Lincolnshire. Binbrook, near Market Rasen, was the location for a number of poltergeist occurrences including the throwing about of objects, spontaneous combustion and animal mutilation.

In January 1905, 226 chickens were slaughtered over several nights at Binbrook Farm. This happened even when the farmer was on guard outside the henhouse. They were all killed in the same macabre way. The skin was torn from around the neck, and the windpipe snapped.

The bizarre death of Snippy the horse on 8 September 1967 in the USA heralded the current interest in animal mutilations. (The name of the mutilated horse was actually Lady, Snippy being her companion, but the initial press reports confused the two and the victim has been known as Snippy ever since.) Snippy lived on a ranch in a remote area of Colorado's San Luis Valley. He went missing and was discovered with his head and neck stripped of skin, his internal organs missing and drained of blood. The cuts were described as carried out with surgical precision. Snippy's death acted as an archetype for the thousands of mutilations which then swept the United States, and later penetrated Canada.

The carnage, which is still happening, was at its worst in the 1970s, when it was estimated that upwards of 10,000 head of cattle had fallen victim. Many were bloodless corpses with internal organs and genitals removed. Some of the surgery was so good that veterinarians at Oklahoma State University stated they were unable to produce students capable of duplicating what the mutilators had done. The ranchers and local law enforcement officers admitted that they were baffled by the lack of footprints, tyre marks and blood even on the ground. No one had heard or seen anyone carrying out the crimes.

There were plenty of witnesses who said they had observed mysterious hovering lights over pastures where mutilated animals had been found. Unmarked helicopters were also sighted near the locations. When air bases were checked out they denied any knowledge of the aircraft. This caused many people to speculate, including television producer Linda Moulton Howe, that the mutilations were carried out by aliens who were extracting genetic material to manufacture EBEs – extraterrestrial biological entities. Indeed, in one Puerto Rican case, a space-suited entity was seen at the location of a mutilation. The lack of blood at the sites

The skeleton of Snippy (actually 'Lady'), a horse who was discovered mutilated in September 1967. Since then, thousands of head of cattle have been found with internal organs surgically removed

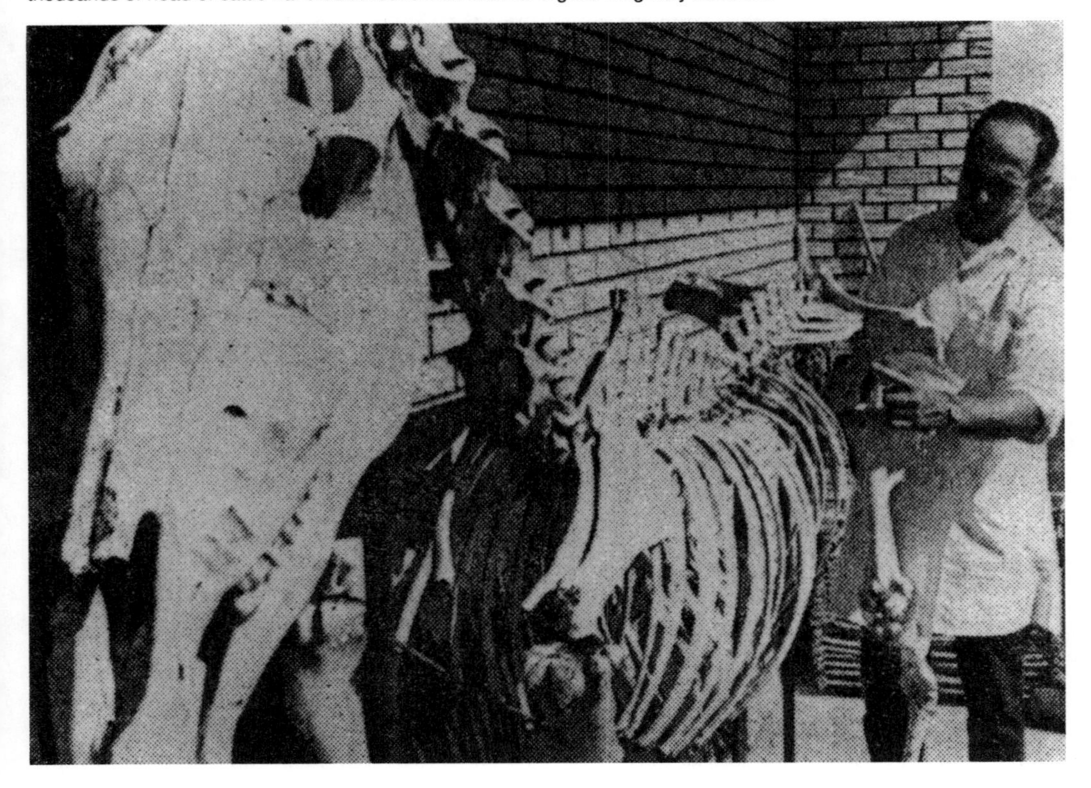

could be explained they said if the animal was abducted and mutilated 'elsewhere', and afterwards returned. This speculation was supported by one case where an animal was found lodged in a tree. Ms Howe made an award-winning documentary on the phenomenon called 'Strange Harvest' originally broadcast in May 1980.

Some critics claimed there was no mutilation mystery at all. The animals had died of natural causes, then their corpses were mutilated by natural predators such as magpies and foxes. Carl Whiteside, Deputy Director of the Colorado Bureau of Investigation agreed with that, and ruled out another scenario which included Satanists. Satanists, it was theorized by some, had been removing sexual and other organs for use in pagan rituals. Whiteside, in a statement made in 1979, could not understand why, if that was the explanation, no one had ever been caught. In any case, the number of mutilations was so large that an army of Satanists would have to have been roaming across thousands of miles of countryside.

Others, including researchers David DeWitt and author George Andrews, believed that the mutilations was being carried out by secret government departments. Scientists claimed that by testing tissue samples from grazing animals, it was possible to detect the presence of oil and mineral deposits. Government agents would then tip off the oil and mineral industry who could purchase the land for very little off ignorant ranchers.

Slaughtered sheep

Mutilations in the 1980s and 1990s have spread to other countries including Britain and Sweden. Charles Fort wrote of sheep being discovered dead with puncture holes in their throats. This caused some talk in the media of 'vampires'. Such incidents are happening again in the UK.

Forty-four ewes were slaughtered in one week during January 1985. They belonged to a farm near Ballymoney, North Antrim, Northern Ireland. Archie Rogers of the Ulster Farmers' Union said it was a mystery how the sheep had died. They had not been mauled, and all of them had puncture wounds to the neck. Dogs were ruled out because there was no evidence of panic among the animals. It was as if they had just laid down and succumbed.

A similar incident occurred at a farm near Rhayader in mid-Wales. The attacks were over several weeks from August to October 1988. Thirty-five sheep were killed, all of them bearing puncture marks to the breast. Peter Hough spoke to the farmer's son, Charles Pugh, who told him, 'The animals were found with four to five small teeth marks. One of them was still alive. It took it two weeks to die. This has been happening just three hundred yards from us, but no one has seen or heard anything. It's the strangest thing we've ever encountered in forty years of sheep farming.'

Mr Pugh invited the local hunt onto their land four times in an effort to track down the beast which was killing their sheep. On each occasion the hounds picked up a scent which led to a local brook. Scientists who examined some of the carcasses ruled out dogs or foxes. Charles Pugh speculated to Peter Hough that it could be an otter with a broken tooth, but he really did not know. The author wrote to the Ministry of Agriculture but his letter was ignored.

Headless seals

The Orkney Islands have been in the news in recent years for allegations of satanic ritual abuse and a devastating oil spillage. But they are also the victim of another scandal – the strange mutilation of otters.

In 1991 thirty seals were discovered on a beach – with their heads removed. Mike Lynch, an inspector with the Scottish Society for the Prevention of Cruelty to Animals said the Society had never come across anything like it before. The heads had not been hacked off, but removed with surgical precision. He added that the seals appeared to previously have been in good health, and referred to the killings as 'a massacre'.

Veterinary surgeons said the cuts were clean, precise and had gone between the bones of the vertebrae without damaging the bones. There was total blood loss, with no evidence of blood at the location. Everyone was baffled.

Horse rippers

Since the late 1980s both horses and cattle have been mutilated in Sweden. By February 1992 around two hundred horses had been attacked. Most of the victims were assaulted in the same manner. They were cut with a scalpel-like knife in or around their sexual organs. Approximately half of the animals had to be destroyed.

No one saw or heard any noise, and there

were instances where even guard dogs failed to react. Both police and owners could not understand how a stranger could get near enough to injure the horses, without them offering resistance. No evidence of drugs was found. A vet stated that the attacker not only had a good knowledge of animal behaviour, but animal anatomy too. Despite the vigilance of police and owners, the sinister attacks continue.

Since the late 1970s, hundreds of horses in the south of England have fallen victim to the phantom ripper. Not all of the attacks have been carried out 'professionally'. Sexual organs have been damaged with blunt instruments in some cases. In one case, which occurred in April 1992, the perpetrator successfully avoided security devices and entered a locked stable. There, a horse was mutilated belonging to Olive and David Gray. The animal had to be put down.

Former sixties pop star, Davy Jones, suffered two attacks on horses at his home in the Meon Valley during July 1992. Genitals were cut, and there was evidence that an object had been forced into the horse's vagina.

Police in Sweden and England have noted that the attacks have taken place at stables close to roads, but as most farms and riding schools are liable to be near roads anyway, this may not be significant.

Once again, it has been speculated that the attacks, many of them on nights of the full moon, have been carried out in connection with satanic rituals. But, again, the perpetrators seem to exercise almost supernatural skill in avoiding being caught – despite the offer of a £1000 reward. Several years ago a man who needed psychiatric help was arrested after confessing to mutilating horses but the attacks continued. In January 1993 a youth was arrested on suspicion, but in March there were further mutilations reported, this time in Cornwall.

The mutilations spread

Animal mutilations are on the increase. There have been cases in many other countries including Denmark and Spain. One of the latest is Japan.

In the early hours of 29 December 1990, a farmer was awoken by the furious barking of his dog. The farmer, who lives near the town of Saga Prefecture, ignored the animal, even though it continued to bark for a long time. At 06.00 hours he discovered the mutilated corpse of a twelve-month-old cow on the floor of the cowshed. Half its tongue was missing, and its four nipples had been cored out from the udder – a feature of many American cases. There was no evidence of a struggle.

When the dog started barking again, two years later on the 4 January 1992, the farmer was not slow this time in investigating. He entered the cowshed and saw a small white object, like a jellyfish, floating in the air. The object drifted outside where it vanished. A cow was discovered lying on the floor. It had a badly broken leg.

Despite the tangible evidence of bizarre animal mutilation, we are no nearer in finding an answer.

A summary of slaughter

- Bizarre animal mutilations have always been with us but not, it seems, on the scale exhibited since the 1960s.
- The sceptics have blamed Satanists and natural predators. Others have sought to blame hypothetical aliens. In truth no one knows the answer.
- Although mutilation includes many animal species it has followed a pattern. In America and Canada it has mainly involved cattle, but in England, Sweden and other European countries, the victims have been mostly horses.
- Most mutilations have been carried out with surgical precision, others display unsophisticated aggression. Are they the same phenomenon, or two entirely different problems?
- Unlike more 'normal' crimes, there is almost a total lack of arrest and prosecution in animal mutilation cases.

MIND MATTERS

THE PSYCHIC DETECTIVES

Newspaper headlines of child-abductions and serial murder cases inevitably attract the attention of seasoned psychics. These people, with encouragement from national newspapers and breakfast television producers, attempt to discover buried bodies or identify psychopathic killers. Sometimes they come up with surprising information, but often, as in the Moors Murders and the Yorkshire Ripper case, many seem way off the mark. Names like Peter Hurkos, Gerard Croiset and the late Doris Stokes, readily spring to mind as psychics who have publicly involved themselves with the investigation of major crimes.

Dutch psychics Peter Hurkos (real name Peter van der Hurk) and the late Gerald Croiset have come in for criticism by the sceptics. Piet Hein Hoebens claimed that checks he carried out on Hurkos showed many of the claims made about him were exaggerated or had no substance at all.

Hoebens admitted that Professor Croiset 'was a genuine challenge to the sceptic', but argued that many of the 'psychic' clues Croiset produced could also have been arrived at through normal logical deductions. Nevertheless Croiset provided many vital clues over the decades, which impressed police officers with their accuracy. Croiset himself admitted that in 90 per cent of criminal incidents he found it

Psychic Gerald Croiset whom Hoebens concluded 'was a genuine challenge to the sceptic'

Clairvoyant Nella Jones has been more successful than most in psychic detective work. The clues she offered on the Yorkshire Ripper murder case proved to have been accurate when eventually Peter Sutcliffe was arrested

difficult to provide enough positive leads to discover the culprit.

The Yorkshire Ripper

Both Gerald Croiset and Doris Stokes were asked to furnish clues to the identity of the Yorkshire Ripper – the brutal slayer of thirteen women. Both psychics came up with details which seemed to be completely wrong. A man named Peter Sutcliffe was finally arrested and charged in January 1981 with the murders.

However, one clairvoyant, Nella Jones, provided some quite accurate information. Freelance journalist Shirley Davenport told Jenny Randles that Nella had provided her with some very accurate clues while the killer was still at large. She knew that the murderer was called 'Peter' and that he worked as a long-distance lorry driver. While other psychics were insisting the Ripper lived in the north-east of England, Nella correctly predicted he lived in Bradford, West Yorkshire, in an elevated house numbered 6. She also predicted that the company who employed him had its name embossed on the cab door and that it began with the letter 'C'. Nella Jones also informed Ms Davenport that the killer would strike once more on 17 November 1980.

The information was amazingly correct. Lorry driver Peter Sutcliffe worked for a road-haulage company called Clark Holdings which was inscribed on the cab door. The Ripper's house was elevated above the road at 6 Garden Lane, Bradford.

A lot of valuable police time was taken up trying to trace the sender of an audio tape of a man boasting he was the killer. It was not Sutcliffe and the police came under a lot of criticism for believing it was while the real murderer roamed free. However, some police officers now believe Sutcliffe had an accomplice, who was probably the man on the tape. Were Gerald Croiset and Doris Stokes tuned into this individual and so not mistaken after all?

Nella Jones also provided clues to the disappearance of a little boy in 1990. Four-year-old Simon Jones was abducted while out playing with his brothers in Hemel Hempstead, Hertfordshire, on 23 September. The case made national headlines. There were no leads, and after a few weeks the police operation was scaled down. The majority of abducted children are usually murdered. Any 'psychic' would have been playing safe by predicting murder in this case. However, Nella Jones told the boy's mother, Sally Jones, that she sensed Simon was still alive. Further, she said a 'bell' and a 'tunnel' were connected with the little boy's whereabouts.

Two months after his disappearance Simon was found in a men's hostel, just a few miles from where he lived. Two women had spotted him there and the police had been informed. A man called Peter May was jailed for twelve years in June 1991. The youngster had not been physically hurt. Interestingly, the hostel was near a pub called The Bell and the entrance was down a narrow alley-like tunnel.

The practicalities of psychic detection work

Behind the big names and the sensational headlines, there are other psychics quietly at work. Dr Keith Wilde is not interested in personal glory, nor the large fees some sections of the media are prepared to pay for 'a good story'. He is a clinical psychologist by profession, with a practice in Manchester, which began in 1958.

In September of that year, Keith Wilde, then a physiotherapist, decided to investigate the claims of people who apparently could bring about paranormal effects. He told Peter Hough what happened. 'Up to that time I had experienced nothing personally. I thought I was as psychic as a brick wall. In fact, I totally disbelieved in it! But I've got the sort of mind that demands answers to any mystery in life. Were these things genuine? If so, could some people be born with something that the rest of us are denied? Or was it something innate that anyone could nurture into life?'

Dr Wilde now believes that psychic ability is something that all of us can learn. Sensitives use psychometry as an aid to pick up information. Psychometry is the ability to gain knowledge about a person or persons through physically holding an object, such as an item of clothing, or a pen, which is associated with them. He tested one medium with a series of objects, and was so impressed that at one stage he wondered if she was picking up the information from his mind. 'I was fascinated, then she said, "One day you'll be able to do this even better than I can." I thought, what a load of rubbish!'

Rubbish or not, Wilde found himself attending a development class when something inexplicable occurred. 'The table I was sitting at suddenly moved and pinned me against the wall!'

He started developing his mind, concentrating for many hours, until he began experiencing strong intuitions about people and places. 'It has more to do with the individual than outside forces. It's the same with hypnosis and to an extent, medical practices. Technically speaking, in my professional work, I've never cured anyone of anything. But I have taught others how to bring about changes within themselves.'

Over the years, Dr Wilde has built up many files on the cases of missing persons he has been involved in. Back in 1971 he was invited to a house in Telford and given some clothing to examine. He said the clothing belonged to a dead teenage girl and even drew the position of how her body lay. The information he passed on to the family was so accurate that several days later he received a call asking him if he would take an all-expenses-paid trip to County Meath in the Republic of Ireland. There, on 12 October, pretty Una Lynsky had disappeared.

He was allowed free rein around the girl's house in a futile effort to pick up impressions which might solve the tragic mystery. Afterwards, the garda drove him to three locations. At the first two he felt nothing at all, but at the third he had a distinct feeling that the abductor had been there. He gave the garda three sets of initials, and immediately an officer recognized one set as matching a man who had been arrested on a burglary charge in that same area some time before.

Dr Wilde told them the body would be found covered with a layer of branches and leaves in a wooded area nearby. The position of the body, he said, was somehow odd, but he could not elaborate. The letter 'W' was also important. It transpired that Una was known at home as 'Winifred'.

The body was found six days later exactly as Dr Wilde had predicted – covered by branches and leaves. It was not lying flat but resting in the lee of a mountain slope. The man who was arrested for the murder had abducted Una while she was out walking, then later dumped the body.

Dr Wilde is, however, sometimes more off the mark. When ten-year-old Jane Taylor vanished in 1966 in Mobberley, Cheshire, Keith Wilde was invited by the girl's family and a local newspaper to try and help. The bike the child had been riding, called 'The Pink Witch', was found abandoned down a country lane.

Dr Wilde arrived at the Taylors' house and under the auspices of Maurice Carver of *The Knutsford Guardian* he successfully picked out some clothing belonging to the girl from items belonging to five other children. In his extensive notes, he wrote: 'I saw a girl with injuries to her forehead lying on the road, opening and closing her eyes as if only semi-conscious. Someone is leaning over her. There's a view through the rear window of a low car . . . '

Jane Taylor's skeleton was found six years later near Colwyn Bay, North Wales. A man in prison on other charges confessed to the murder to a fellow prisoner. He had followed the girl down the lane in his Mini car, waited for her, then thrown a rope around her neck as she cycled past.

Keith Wilde had pin-pointed a pub in the village which he felt possessed a strong connection with the disappearance. During the trial it was ascertained that the murderer had frequented that pub regularly. But he had also spoken of the killing as being 'an accident' and gave the police a car registration number which made no sense. 'These wrong results interest me more than when I'm right,' he told Peter Hough. 'Why was I wrong, I ask myself? I believe the actual information is never wrong, but it arrives in some sort of code. It is the misinterpretation of this code which leads to incorrect conclusions.' Yet despite this, sometimes the feelings are so strong that Dr Wilde's convictions override the strong opinions of others.

When Staffordshire girl, Christine Darby disappeared in 1967, police built up an Identikit picture of the killer. But the face which burned in Keith Wilde's mind was that of an older man wearing glasses. He became so obsessive with this image that he took to carrying around a camera until he saw a man who matched the description. This man, of course, had nothing to do with the crime, but now Keith had something tangible to show the police.

Eventually a man was arrested who resembled the police Identikit picture. Was Dr Wilde completely wrong? Although the police description matched the killer, Keith Wilde's picture did in fact resemble that of the killer's father. It transpired in court, that the father had been lying to the police concerning the whereabouts of his son.

'Why should I pick up *this* man?' Dr Wilde asked himself. 'Emotion is the registering force in all this. Imagine the internal trauma of a father lying to protect his son suspected of murder. When I started looking for a lead, it was this man's emotional discharge I picked up!'

Psychic detection in the modern world

Michael Bromley is a Celtic shaman. This burly six-footer, bearded, with shoulder-length hair, claims he can commune with the spirit world. He spent seven years in America working with native Indians, searching for missing persons and psyching out security loopholes at the Los Angeles Olympic Games.

'Shaman' is a very old word which originated in Siberia, although it is usually associated with the Red Indian medicine man. It is based on the ancient belief that there are many levels of 'spirit' both in the spirit dimension and on the Earth plane.

Michael Bromley, a celtic shaman, assisted police officers at the Los Angeles Olympic Games, psyching out trouble spots

In 1984 he was approached by organizers of the Los Angeles Olympic Games to help them search out security problem areas. He met Patrick Connolly, Los Angeles Chief of Police, and they sat down and talked. Michael was assigned a police officer who was instructed to drive him around the vast Olympic site. After spending weeks visiting various locations and concentrating on maps of the area, he compiled a report.

'It was no good using airy fairy language,' he told Peter Hough. 'This was for the corporate mind. I had to be specific and included many photographs. It had to be the sort of thing a tough-nosed cop could read and understand. What I didn't know, at the time, was that copies were sent to the CIA, the FBI and five other security forces.

Michael claimed he could 'sense' at a specific location when there was the probability of trouble. Westwood, he wrote in his report, was an area of tremendous negativity. On the opening day of the Games, a young man drove down the main street and deliberately ran over about twenty people. At his trial, he said he felt 'waves of energy' hitting him. After it was all over, Michael sat down with Pat Connolly to determine how accurate his dossier had been.

'I had predicted low, medium and high risk days. I was 100 per cent on that. There was some terrible violence on the high-risk days, although I couldn't always predict where. But I was right about some shootings on one of the freeways. I also accurately predicted that a private security guard would attempt to rape one of the women athletes. I actually predicted the area. There was even a bomb hoax at LA Airport, although it occurred two days after the games!'

Afterwards, Chief of Police, Pat Connolly commented, 'Mr Bromley pin-pointed certain problem areas with more than a high degree of reliability.'

But what of the 'waves of energy' described by the young man. Was this merely indicative of a disturbed mind? Michael thinks differently.

I lived outside of LA. Quite independently, people were calling me from the city saying they felt waves of energy coming into the area. The Soviets didn't come to the Games. All the phone lines from that country to America were engaged a lot of the time. I realized that phone lines were carriers of energy. Now if it was possible to send psychic *energy down those lines . . . I believe the Soviets were projecting negative energy into Los Angeles. I know it sounds far-fetched, but we all do it every day. It's the same as wishing someone well, sending our love, or wishing them harm. The Russians have been carrying out scientific experiments into parapsychology for decades.*

Michael Bromley had similar successes with the 1986 San Fransisco Bay Olympic Games, and the Seoul Olympic Games in 1988, as South Korean government minister, Ri Hoon Hur, verified.

The publicity, particularly from his achievements in the Los Angeles Games, attracted individuals with their own unique problems. One of these was the wife of a man who had gone missing on 1 November 1985, in Mariposa County. The elderly gentleman had set out for an afternoon walk and failed to return. The police were inclined to believe the man had just run away but his wife feared the worst. Tom Strickland, the Acting Sheriff-Coroner, made these comments, 'An extensive search was started by the Mariposa County Sheriff's Department using search dogs, helicopters, 4 wheel-drive vehicles and horses. After nine days the search had to be suspended due to heavy snowfall, and the difficulty of the terrain. At that time not a trace of the gentleman had been found.'

Michael Bromley contacted the Sherrif who invited him to assist. 'I meditated on the man's life-force but could detect nothing at all. I was convinced he had been murdered. I pinpointed the area where the body lay, and said it would be discovered by two people. I was right on both counts.'

Tom Strickland verified this: 'On 9 February 1986, the remains of the gentleman were found by prospectors. The remains were out of the search area established and was in the direction described by Michael Bromley. He had also given us many new leads, and had been extremely accurate on the conclusion of the search for the missing man.'

Michael added, 'I also gave them a description of the killers, but they wrote it off as misadventure. He was an old man, there was little chance of getting the murderers I suppose.'

Later that year, the *Mariposa Guide* called him in to test his psychic abilities on an eighteen-month-old unsolved double murder. Inez and Robert Roos had been discovered shot dead. Michael told Peter Hough he had come up with some very concrete information: 'I told them it was a Mafia killing. The man gambled heavily and owed money. The assassins came and ordered him to pay up or they were prepared to kill him. He thought they were joking. They took his wife outside and shot her, then they killed him.'

He gave a lot of new information to the police, including details of a corrupt bank employee, but the police, who had already spent hundreds of manpower hours on the case decided to drop it, as tangible proof was not forthcoming. In conclusion, the Senior Editor of the *Mariposa Guide*, made these comments: 'Mr Bromley did pin-point certain questions, offering a surprising amount of original information, giving new leads and direction to the investigations.'

The danger of being too accurate

Most police officers would react with suspicion to someone offering 'inside' knowledge of a murder. They would more likely believe that person was connected with the crime than accept the notion of a paranormal explanation. That is exactly what happened when Etta Smith offered some very accurate information to police which led to the discovery of the body of a murdered Californian woman.

On 15 December 1980, a thirty-one-year-old nurse called Melanie Uribe disappeared as she drove to work through Los Angeles. Etta Smith first heard about the disappearance from a friend in the nursing profession, then in detail on a news broadcast. As she listened to the report Etta was overcome by strange feelings and 'knew' the missing woman was lying dead in a canyon. She talked to Lee Ryan, one of the detectives investigating the mystery, and pinpointed Lopez Canyon as the location, and gave a description of the clothes the victim was wearing.

Clues to psychic detection

- On occasions so-called psychics can tune into information which it is reasonable to conclude is more than just coincidence, or the normal deductive powers of the mind.
- Psychics claim that information arrives in code, and it is a failure on their part to misinterpret the code leading to wrong conclusions.
- Far too often, however, exaggerated claims are made in order to bolster reputations and create a good story.
- Many senior police officers are open minded enough to include 'psychics' in their investigations. If it was all guesswork interest in psychic detection would have waned long ago.

Back home her children persuaded her to drive out to the area to see if they could find anything. They did – there was something white sticking out of the dust. On the way out of the canyon she flagged down a police car and led them back to the spot. It was indeed the body of Melanie Uribe.

The police could not accept that Etta could have known where the body was, and also give an accurate description of the woman's clothing unless she had been involved in the crime. They questioned her in a police cell for three days, deciding whether to charge her or not. But finally they arrested three men who were convicted of the rape and murder of the nurse.

In March 1987 Etta took the Los Angeles police to court for wrongful arrest. She was awarded damages of $26,184 – about £16,000.

THE PSYCHIC SURGEONS

I was in Baguio [in the Philippines] from 2nd to 19th of November 1980. My reason for going was that I was suffering from cancer of the prostate gland which had disseminated to the bone. This had been diagnosed in Europe and I was hospitalized in Bregenz for three weeks before leaving for the Philippines. I was operated on five times, was treated with infusions of cytostatic agents and had various brain scans to see if the metastases had spread to other parts of my body.

In spite of the operations, the removal of testicles and the infusions, the pain was almost unbearable. I could hardly lift my right arm. This was the condition I was in when I reached Baguio.

In the course of the following sixteen days I was operated on by Placido twenty-eight times.

Rustico treated me daily, touching and massaging the painful areas. Josephine operated once on my shoulder, to where the mestases had spread. I went to the healing centre and Teodor treated me with his 'aparatos'. Flores gave me his spiritual injections, and Alex and his team massaged my body.

Right now I feel fine. I am convinced that the healers' treatment and that of Thomas Schuller, a homoeopath, are responsible for this improvement. I have no pain whatsoever and have gained 7 lb in weight. I have started working again full time. When I returned to Europe everyone was mystified. None of my doctors could believe that the cure might be permanent.

The above letter is typical of the many received by Gert Chesi, investigator of the psychic surgery phenomenon. This controversial subject came in for popular debate in the 1960s when Westerners returned from Brazil and the Philippine Islands with amazing stories of miraculous cures. They described surgery conducted by uneducated peasants who opened bodies without pain and plunged unsterilized hands inside, extracting diseased fibrous tissues.

By the 1970s these so-called 'bloody operations' had taken the public imagination by storm, prompting many investigations in which the healers usually were only too pleased to participate. The enigma would soon be solved, it was thought. Either the phenomenon was authentic or it was a cruel confidence trick

In the Philippines, so-called psychic surgery was at its height in the 1970s. Practitioners claimed they could open a body with their hands and painlessly extract tumours. Was it all a fraud?

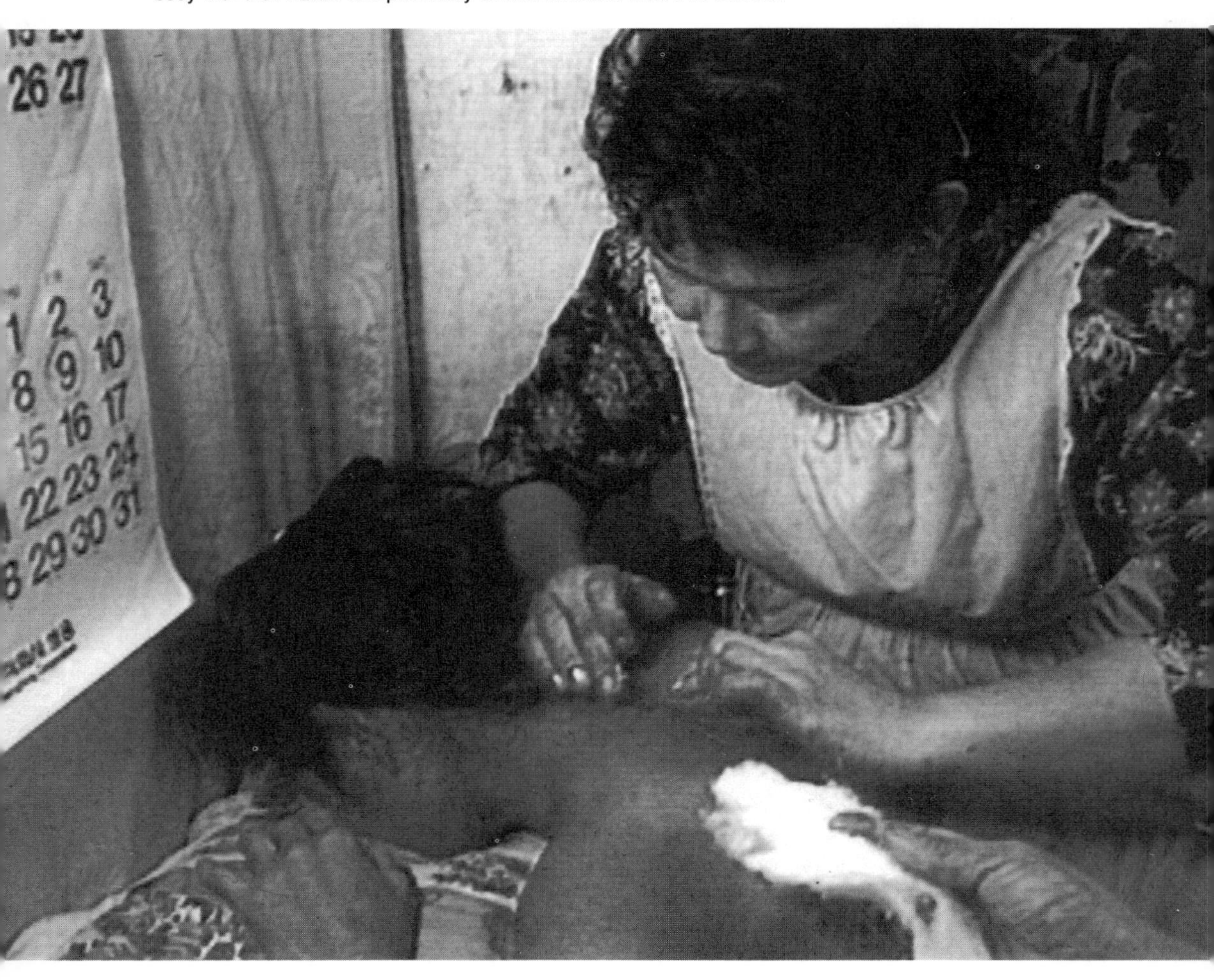

A crop circle formation at Alton Barnes, Wiltshire, discovered in July 1990. Although crop 'pictograms' such as these now appear to be hoaxes, simple circle formations have been around for many centuries, discrediting the idea that all crop circles are the result of modern practical jokes.

Photographed by Ella Louise Fortune on 16 October 1957 over Holloman Test Range, New Mexico. Visually the object had a clearly defined edge. Although some analysis has indicated it might have been an unusual noctilucent cloud, this may depict the plasma type of UFO.

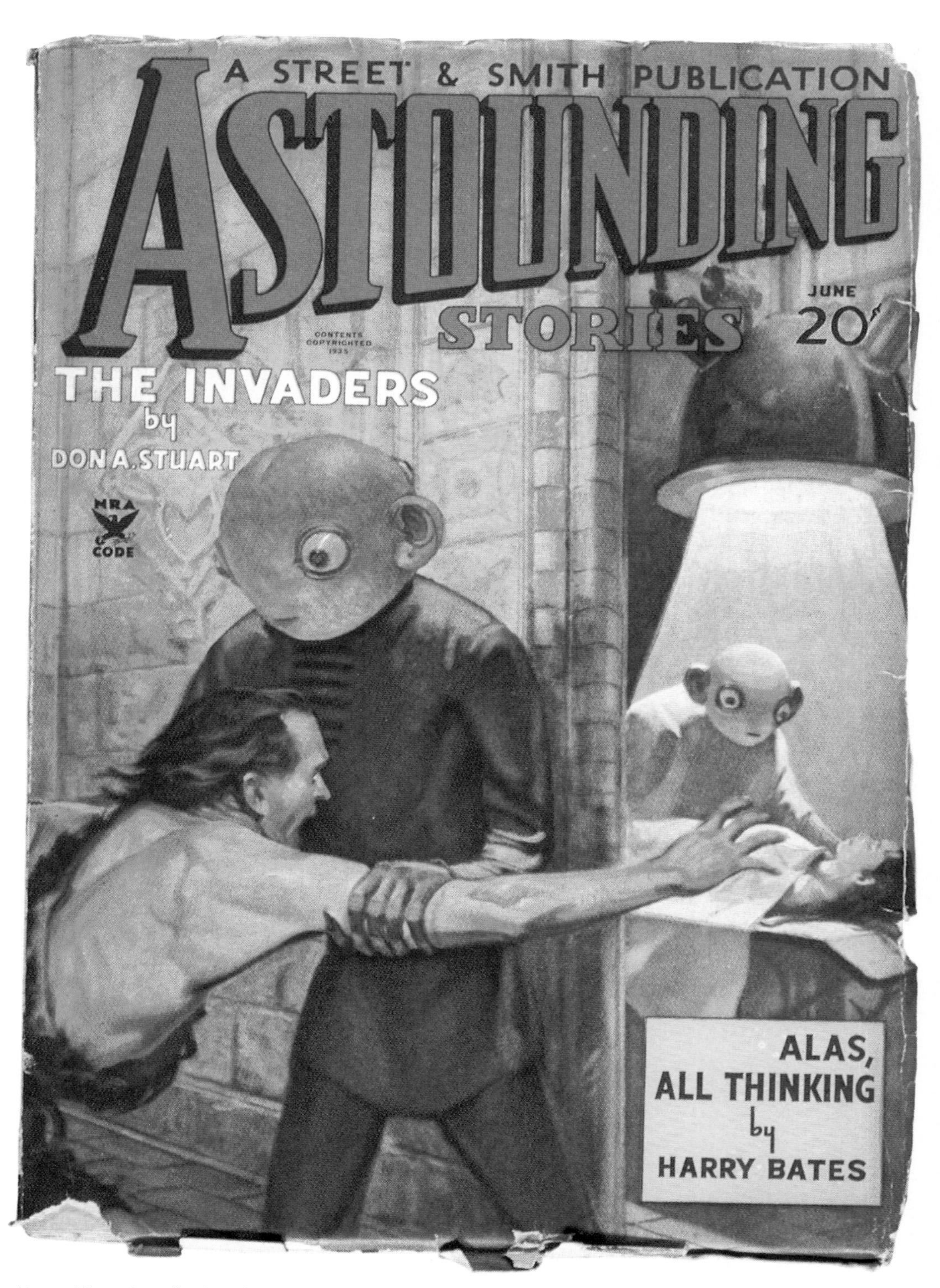

Howard Brown's realization of a tale by John Campbell (writing as Don A. Stuart) in *Astounding Stories*, June 1935. The scene shows some similarities to modern abduction accounts 22 years before the Antonio Villas Boas case in 1957, the first recorded case of alleged alien abduction.

Above Fireman Tony McMunn's inauguration into the spontaneous human combustion controversy was a true baptism by fire. On 4 March 1980, he was confronted with the calcined remains of an elderly lady. The upper part of the body was pure ash but, strangely, the lower legs and feet were intact and there was no fire damage to the surrounding area.

Left Deputy Chief Fire Officer for Hereford and Worcester, Tony McMunn. Tony believes in *preter-natural* combustion – that there has to be a source of ignition before conflagration can begin.

A possible ball lightning effect photographed in the summer of 1978 by Werner Burger, at Sankt Gallenkirch, Vorarlberg, Austria. Ball lightning was confined to mythology by scientists who refused to believe eye-witness accounts. Now many atmospheric physicists are exploring the possibilities.

In 1978, the day after his mother's death, Swiss metal-bender Silvio photographed a luminous ball. The picture showed an angel-like figure inside a yellow circle. Did a plasma ball interact with Silvio's thoughts, or is it proof of angelic beings?

The famous visionary, Bernadette Sourbirous, who perceived the image of the Virgin Mary at Lourdes in 1858. Since then there have been many more cases. Religious belief often seems to shape paranormal events to reflect and reinforce existing dogma.

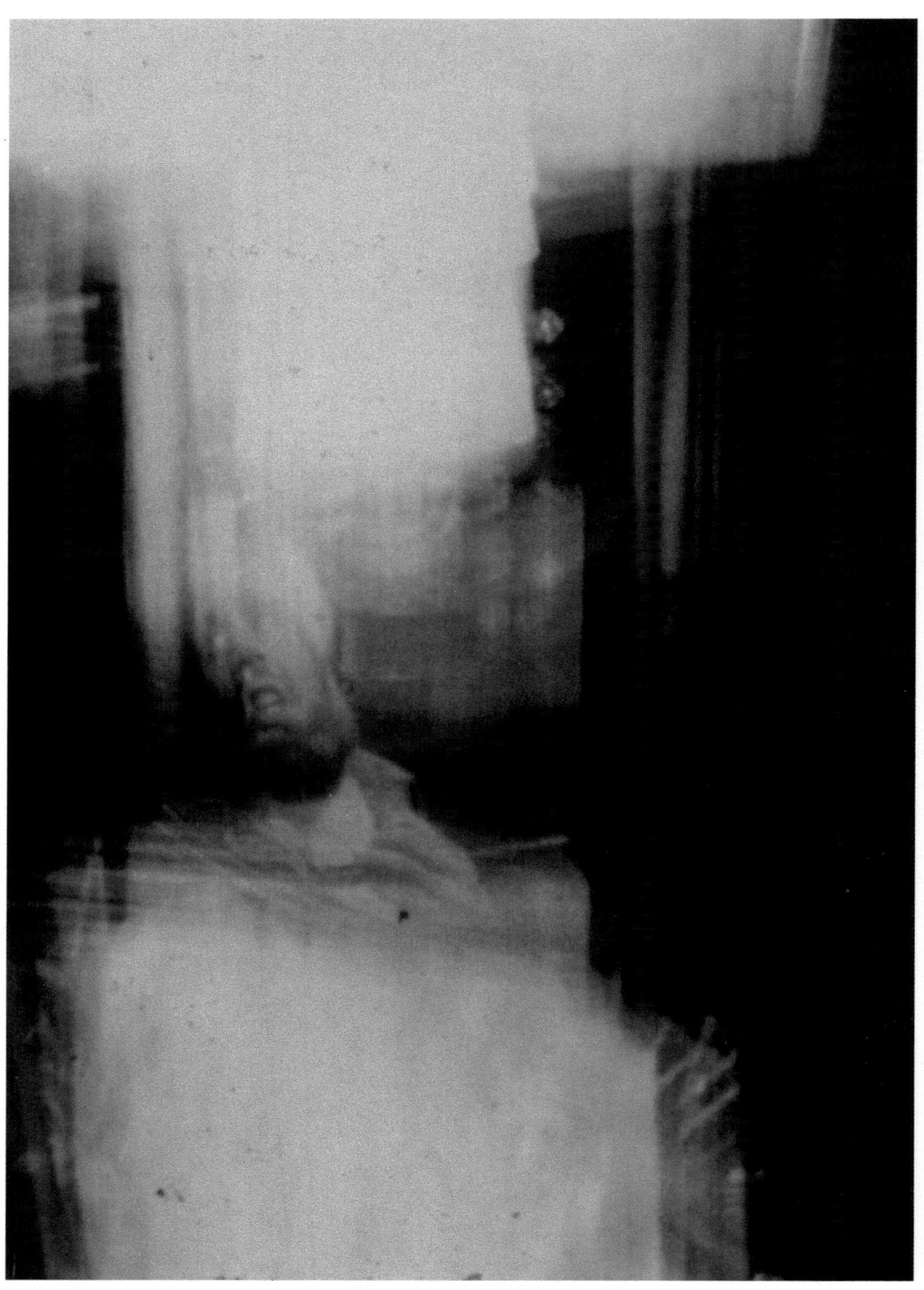

One of many anomalous images to turn up on cine film used by Stella Lansing. This picture of a turbaned man holding a 'flute' appeared on film which should have shown a television programme. Mrs Lansing has been extensively investigated by Dr Berthold E. Schwarz who sees a correlation between psychic and UFO phenomena.

An artist's conception of the sinister Men In Black – bogus officials who plague UFO percipients. These smartly dressed figures seem to appear out of nowhere. Evidence of the 'reality' of the MIB is often provided by the statements of several witnesses. In recent years reports of phantom social workers have paralleled those of the MIB.

A 'sea monster' photographed by Robert Le Serrec in Stonehaven Bay, Hook Island, Australia, in 1964. Critics have suggested it might be weed on the shallow sea bed. Nevertheless, the number of other sightings of alleged 'sea monsters' indicate a genuine mystery.

Psychic surgery was not an invention of the 20th century. Swiss physician and philosopher, Paracelsus, born in 1493, wrote in his *Philosophia Sagax*: 'He who reaches into a man without harming him, without opening him, is as one who reaches into water to take out a fish without making a hole in the water'

played on the gullible desperately searching for wondrous cures to afflictions spurned by conventional medicine.

Although certain television documentaries presented persuasive evidence of fraud, other investigators were not so convinced.

It is interesting, and perhaps significant, that psychic surgery, as opposed to faith healing, which is the laying on of hands, centred in only Brazil and the Philippines. Both are poor countries with social inequalities, have high birth rates and are nominally Roman Catholic. Separated by thousands of miles of ocean, each is situated on similar lines of latitude – one just north and the other just south of the Equator. Both have strong spiritist movements – with a belief in communication with spirits.

The Philippine surgeons

The Philippines cover an area similar to Great Britain. They captured the world headlines in recent years with political and environmental upheavals that saw the expulsion of their super-rich leaders, the Marcos, and the eruption of Mt Pinatubo in June 1991.

It seems that the first public demonstration of the 'bloody operation' was performed in 1945 by an old faith-healer called Elueterio Terte. The principle was then passed on to other healers. However, the alleged paranormal opening up of bodies had been evident in other parts of the world centuries before. The maverick physician and philosopher, Paracelsus, born in Einsiedeln, Switzerland in 1493, had this to say in his *Philosophia Sagax*: 'He who reaches into a man without harming him, without opening him, is as one who reaches into water to take out a fish without making a hole in the water. He can thus reach into a body, practice the fourth specie of *Nigromantie*, that means he can reach into a body and put something in. This we call *Clausura Nigromantia*.'

Gert Chesi, an Austrian paranormal investigator, made two trips to the Philippines in the autumn of 1980 and in February 1981. The results of his many interviews and observations were published in a book called *Faith Healers In The Philippines*. There he witnessed and photographed many 'bloody operations'. While at the small surgery of Marcelino Asuigui he watched the healer operate on a young woman from Australia, who had complained of continuous lethargy 'The healer's hands move over her stomach,' he wrote. 'His fingers pierce through the skin. He searches inside and pulls out a large worm, at least 20 centimetres long.'

Marcelino dropped the worm into a bowl of water, then produced two stones, an egg-shaped organic looking object and two more worms. At least one of the worms was alive. However, when asked to explain the production of the first worm, Marcelino tellingly explained: 'I did not look for it. I suddenly held it in my hands. It is difficult to explain.'

Typical of most of the healers, Marcelino Asuigui has very little medical knowledge. Many healers claim to have no powers themselves, but are channels for spirits, or even God. It does seem that the less the healer understands of physiology, the less the extracted objects resemble genuine tumours. In extreme cases some productions are totally alien to the human body. Is this because the operations are fraudulent, or is there a paranormal explanation?

Juanito Flores was approached by a female patient complaining of a metallic taste. He laughed and told her she probably had 'money in her belly'. This was no flippant remark. Flores made an opening on the right side of her body, and slowly extracted three American quarters dated 1965, 1967 and 1970. He plunged his hand in again, eventually producing twenty metal staples. The operation caused a great deal of discomfort but the wound was closed without leaving a scar.

What did this mean? Was the patient really carrying all these metallic items in her body? Even if she was, would they give her a metallic taste in her mouth?

Even more bizarre was the psychic surgery practised by Rosita Agaid. She extracted palm branches, leaves, nails, string and other foreign objects. Gert Chesi recorded his bemusement when confronted with such irrational phenomena. 'Did this leaf ever hang on a tree? Is this nail the work of man, and if so was it taken from our material world by magic to be materialized later through the power of thought? Are these processes subject to our earthly conception of time, or am I the victim of a clever trick?'

Many healers explain such objects as materializations or apports – objects not really

Paranormal investigator, Andrija Puharich, experienced psychic surgery first-hand when José Arigo put him under the knife – literally. Arigo used a penknife to remove a benign tumour from Puharich's arm. Puharich became convinced that an 'intelligence' was driving the Brazilian

José Arigo on trial in 1964. He was imprisoned for eight months for practising medicine illegally. During this period he was investigated by the authorities, and Judge Filippe Immesi witnessed for himself a successful operation by Arigo

produced from inside the body at all, but manifesting at the psychic openings. They are seen as being symbolic of the patient's complaint.

Brazilian psychic surgery

'Brazil is the most psychically oriented country on Earth', according to Guy Lyon Playfair, the paranormal writer and researcher. Similar to the Philippines, its people are very spiritually aware, readily accepting with little argument the 'reality' of paranormal phenomena.

Investigations into psychic healing began in the 1960s by noted American researcher, Dr Andrija Puharich. Although some of the Brazilians use just their bare hands, others are renowned for using surgical instruments, household cutlery, nail scissors and even penknives. Brazilians, like Filipinos, are generally ignorant of anatomy, surgical technique, and utterly disregard hygiene and anaesthetics.

One of the most widely investigated healers was an iron-ore miner called José Arigo, born on 18 October 1921 into a family of ten. Arigo claimed his bizarre talent first emerged in an incident which occurred around 1956. He was with relatives and friends of a woman dying from cancer of the uterus. As the final minutes of the woman's life ticked away, Arigo suddenly rushed from the room to reappear moments later carrying a large kitchen knife. After ordering the astonished crowd to stand back, he pulled the bed sheets aside then plunged the blade into her vagina. Arigo violently twisted the knife about, removed it and inserted his hand, producing a huge tumour the size of a grapefruit. The patient had felt no pain during the operation, and later a doctor certified the growth was indeed a uterine tumour.

José Arigo was as amazed as everyone else. The woman recovered completely. Arigo had no conscious memory of the act for he had been in a trance. It later transpired that as a child he had experienced visions and heard voices speaking in what sounded like German. He remembered the materialization of 'a monster with a huge belly and German looking face' who later turned out to be a Dr Fritz. It was Adolph Fritz,

Arigo claimed, who was responsible for the amazing results.

Arigo remained in good favour with the Roman Catholic Church. Father Virgilio, parish priest of Congonhas, remarked, 'How can we condemn him for wishing his neighbours well?' Unfortunately, Arigo did not remain on such good terms with the law. In 1958 the Minas Gerais Medical Association managed to have him prosecuted for 'illegal practice of medicine'. He was sentenced to thirty months, but a public outcry persuaded the President, Juscelino Kubitschek, to personally intervene and set Arigo free.

In 1964 he was charged again and served a total of eight months in jail. During this period he was investigated by the authorities, and Arigo invited Judge Filippe Immesi to observe one of his operations. The Judge agreed to hold the head steady of a woman almost blind with cataracts, while Arigo attended to her. John G. Fuller records the judge's testimony in his book, *Arigo, Surgeon Of The Rusty Knife.*

> *I saw him pick up what looked like a pair of nail scissors. He wiped them on his sports shirt and used no disinfectant. Then I saw him cut straight into the cornea of the patient's eye. She did not blench, although fully conscious. The cataract was out in a matter of seconds. The district attorney and I were speechless, amazed. Then Arigo said some kind of prayer as he held a piece of cotton in his hand. A few drops of liquid suddenly appeared on the cotton and he wiped the woman's eye with it. We saw this at close range. She was cured.*

Andrija Puharich sampled Arigo's skills first hand when the healer removed a benign tumour from his arm in seconds. This was done with a penknife which caused no pain. An interview with Puharich was published in the October 1973 issue of an American magazine called *Psychic.* He wrote of Arigo:

> *In the case of someone like Arigo who not only said things like precision medical diagnosis, but who did things – operated on living tissue, made tumours dissolve and leukemia disappear – there was intelligence in operation and action. Even though I couldn't prove it, I was convinced that everything that Arigo did was based on the action of some outside intelligence. I know that there is intelligence involved. And I know that there is energy involved – material systems are influenced, transformed, deformed – I mean things vanish completely and never come back. So energy is being manipulated. My basic hypothesis – and this is just guesswork – is that the fundamental mechanism at work behind it all is the phenomenon of dematerialization and rematerialization. I think that the actual energies at work are going to be found in known physics. I think we're going to actually uncover a fifth force in nature.*

José Arigo died at the age of forty-nine in a car crash during January 1971. It is thought a heart attack caused him to lose control of the vehicle which left the highway and crashed into a tree. Apparently he had warned his friends in advance that something was about to happen to him.

Antonio Sales also began his career during the 1950s. A bricklayer by trade, he looked every bit the conventional surgeon with his white coat, shelves of medicines and surgical equipment. His patients remarked on his pleasant bedside manner. Like Arigo, Sales also claimed to be manipulated by the enigmatic Dr Fritz.

His operations parodied normal surgical procedures, but Sales lacked even a formal education, ignoring sterilization, anaesthetics and working much quicker than conventional surgeons. Guy Lyon Playfair spent two years in Brazil and recorded this statement from a witness to an operation to cure tonsillitis, 'He thrust a pair of scissors into her mouth and started cutting away, telling the girl not to worry because everything was sterilized. I noticed a smell of ether but saw no sign of anyone sterilizing anything. Some blood came out of the girl's mouth, but very little, and two hours later she was eating a huge steak.'

In common with Arigo, Sales wrote out a prescription for huge quantities of antibiotics which completed the treatment.

Playfair, in his book, *The Flying Cow,* told of his own treatment by healer Edivaldo: 'Edivaldo's hands seemed to find what they were looking for, the thumbs pressed down hard, and I felt wet all over, as if I was bleeding to death. I could feel a sort of tickling inside, but no pain at all. The most unusual sensation was a sudden strong smell of ether, which seemed to come from my stomach area.'

At the end of the treatment, Edivaldo muttered something, slapped a bandage on the stomach, then a female assistant told the investigator he could go home. Playfair, however, experienced great difficulty standing up, experiencing the sensation of a hole in his stomach. A taxi was finally called to transport him. Another operation followed with frequent check-ups. He claimed to have been completely cured of a digestive problem for over a year.

Psychic fakery

A television documentary in the late 1970s was made in the Philippines. The crew were invited to film some of the 'bloody operations', and interviewed several of the patients who had flown in from Britain. The evidence they produced seemed damning.

The healers usually 'operated' on fleshy parts of the body, the stomach for instance, because, according to the programme-makers, it was easy to make it appear that the hand was penetrating the abdomen. Animal tissue and small sacs of blood were transported to the area of the operation by the use of sleight of hand. Some of the material which was later analysed showed that it was probably of bovine origin.

When considering charlatanism, one looks for a reason, and that reason is usually the acquisition of money. Chesi found that most healers charged nothing for their services, or just asked for donations. It was believed by them that a lust for material things meant a loss of the healing gift. Most too were very religious, although it mattered little if their patients were followers of Christ, Mohammed or were atheists. Some, however, had no scruples about capitalizing on their 'gifts'. They charged exact fees, bought opulent houses and drove big American cars. A case in point is that of Tony Agpaoa.

This articulate Filipino was one of the most publicized practitioners. He was originally investigated in 1966 by veteran researcher Harold Sherman, who concluded he was authentic. Agpaoa featured in many books, magazines and documentaries, and even published his own booklet modestly titled, *The Living Legend.*

Many became convinced that Agpaoa was less than genuine. Gert Chesi was treated almost with contempt when seeking his cooperation during research for his book. In Baguio, Chesi was told by many people that the Filipino had lost his healing powers years ago, and now faked the operations or relied on the abilities of his assistants. Of course, such remarks could have been more to do with professional jealousy. Speaking generally, healer Eduardo Simbol commented, 'Fake healers pretend to be healers in order to extract money from people who seek help, and this naturally damages our reputation. These fakes should be rooted out mercilessly.'

Gert Chesi made two trips to the Philippines. At the end of the first he was convinced that the 'bloody operations' were by and large authentic, but the second visit deflated that conviction. He admitted, 'It is no easy matter to determine precisely where authenticity ends and fakery begins. At this point I am almost fully convinced that these operations are rarely genuine, if ever.'

Even though he was present at hundreds of operations and detected nothing overtly fraudulent, Chesi became certain that sleight of hand played its part. This view was reinforced by healer Arsenia de la Cruz, daughter of the original psychic surgeon, Elueterio Terte. In her opinion 90 per cent of all operations were faked, a mere concession to the patient who expected to be operated on. This did not mean the patient had not been cured for many testified they had, including their doctors but that the operations themselves were clever pieces of stagecraft. Many Filipinos practice 'magnetic healing', similar to the 'laying on of hands' in the West.

Filipino healer, Teodor admitted, 'I am primarily a magnetic masseur. I do use tricks sometimes though. Whenever I cannot feel the vibrations or realize that the patient does not trust me, I use my "apparatus".'

Arsenia de la Cruz demonstrated how easy it was to palm a small bag containing blood to the operation area, then burst it by kneeding it into a fleshy part of the body. The evidence is discretely removed in the wad of cotton wool used to clean the 'wound'. The superfluity of the operation ties in with comments made to Guy Lyon Playfair during an interview with Edivaldo. 'The patient was already being operated on as he waited in line, and the part where he lay on the bed was only the end of the process. The actual manipulation was only to convince the patient that he really had been operated on. How would the client feel if he got onto the bed and was told the operation was over?'

Edivaldo was not claiming his healing powers were fraudulent, but that the operations were merely a piece of unnecessary window-dressing. In his final days, Arigo confined himself to diagnosis and prescription, such was the faith in

Surgical cuts

- Most of the 'bloody operations' in the Philippines have been demonstrated to be fraudulent
- Some practitioners of psychic surgery explain that the 'operations' are not really necessary but are window-dressing while the real healing is done with the mind.
- Other psychic healers claim that the objects apparently removed from the abdomen materialize in their hands and are symbolic of the illness.
- Enough psychic operations have been witnessed to verify their authenticity. Patients are cut using kitchen knives without anaesthetic or sterilization, by medically ignorant healers.

his powers. It is estimated that he treated two million people in his lifetime. Could a stage magician perform the same trick two million times without being found out – even once?

Psychic surgery in Britain

Psychic surgery is being practised in Britain by a former carpenter called Stephen Turoff. He and his assistant wife have allegedly been carrying out near-miracle cures from their home in Danbury, Essex.

Turoff started out healing by the laying on of hands, which grew so hot they would occasionally burn patients. Later he became controlled by the spirit of a dead German called Dr Kahn, sometimes assisted by a Mr Jones – Turoff's spirit guide. This was when he began carrying out surgery, with no anaesthetic, no disinfectant, with the aid of a penknife. As with other psychic surgeons the results are varied and inconclusive – at least on a scientific basis.

Mr Turoff seems to have the knack of correctly diagnosing illnesses that patients have not made him aware of. At one session he 'operated' on a Mr Renford Stephen for back pains. In a trance, Dr Kahn took control and announced in a German accent that Mr Stephen also had sinus trouble. Kahn – or Turoff – then went through the motions of using a hypodermic syringe to administer anaesthetic, and the patient reacted to the non-existent jab. The penknife was then inserted about an inch into the patient's nostrils. Afterwards Stephen said he had felt no discomfort and claimed that the sinus trouble had gone.

Next, Turoff treated a reporter from the *News of the World* who complained of back pains. More seriously, the journalist had suffered bone cancer for three years, but kept this from the healer. Turoff as Dr Kahn, touched his head, arms and legs, and decided an operation was not necessary. Instead he placed his hands on the reporter's back for four minutes, who said afterwards, 'The heat from his hands was incredible. It was as if he'd placed a hot iron against me. I felt instant relief.'

Next, Turoff correctly touched the areas which were cancerous, and the man felt an immediate improvement. At the time he was having chemotherapy, and he found during subsequent sessions with the healer, that even when the pain was excruciating, there was always immediate relief.

Edward Harris, editor of the journal *Cosmology Newslink*, visited Stephen Turoff in April 1990 with a long-standing back problem. True to form 'Dr Kahn' also discovered another illness, and promptly used a scalpel on Harris's stomach. As far as the back pain was concerned, it was much reduced, but not totally cured. However, Turoff told him the complaint would need another visit, which did not take place.

He charges a nominal fee for each session, but claims that if the sick cannot afford to pay he treats them anyway.

Turoff said that sometimes blood was produced during the operations, and told reporters, 'I am scared stiff. I don't know why I have been singled out for this gift. I have no recollection of anything afterwards.'

Some critics conclude that many of these 'miraculous' cures are psychosomatic. If this is the answer, let us see millions of pounds poured into research. A cure for cancer using the power of the mind has tremendous application.

Dr Hans Naegeli Osjord and Dr Alfred Stelter agree that a cure rate of about 30 per cent occurs in the Philippines. This may seem low for a miracle, but remember that people who turn to psychic surgery for help are hopeless cases. These are desperate people for whom orthodox medicine has failed. Saving one out of three condemned to die is highly remarkable.

THOUGHTOGRAPHY

Can images be impressed onto photographic film by the conscious or unconscious will of the mind? There are many examples of anomalous marks and shapes turning up on family snap shots. Peter Hough investigated the claims of a Lancashire woman, under the auspices of ASSAP (Association for the Scientific Study of Anomalous Phenomena).

Mrs Rita Smethurst first contacted ASSAP in July 1990. She explained that over the previous ten years white blobs and streaks of light had sometimes appeared on photographs taken during family gatherings. In a subsequent letter, Mrs Smethurst explained that she took 'hundreds of snaps in a year', and it was only on a few of them that the anomalies appeared. These varied from a grey shadow on a photograph taken at her daughter's wedding, photographs taken at her brother's farm showing a white light and black shadow respectively, and anomalies appearing on nine photographs connected with her son's wedding in the summer of 1990. The most remarkable of these depicted a blue hose-like effect which manifested on two consecutive pictures of the wedding cake and wedding dress taken at the beginning of July.

Peter Hough visited Mrs Smethurst and discovered that she had used two different cameras which had both produced anomalies. They were an Olympus Trip AF MO and Olympus Trip S model. On another occasion she had used her son's camera. This seemed to rule out malfunctions connected with a single camera. A selection of prints and negatives were sent to a photographic analyst, Vernon Harrison, by Michael Lewis, National Investigations Coordinator for ASSAP. Mr Harrison agreed with Hough's conclusions: 'I do not think that the anomalies noted are caused by double exposures, leakage of light into the film cassette or through the camera back or shutter, processing faults, bad prints made from negatives or faulty operation of the camera shutter.'

Vernon remarked that the black shadow effects were not normal shadows cast by material objects, but regions where the image had not recorded properly. 'One could get this effect if a dark, opaque object such as a twig were interposed in front of the camera lens. It seems rather unlikely that this could have happened without Mrs Smethurst noticing.'

Similar explanations including camera straps and stray fingers partially blocking the lenses, were also suggested. Two photographs taken on 5 April 1991 at a local beauty spot are a case in point. Both exposures show a fuzzy black line running across the bottom right-hand corner. Yet if one studies the anomaly, trees and daffodils can be seen through the image.

Harrison commented of the two pictures showing the hose-effect, 'What is particularly interesting is the presence of striations which are so regular as to suggest the flexible tubing attached to a vacuum cleaner. I have had similar examples shown to me by the SPR [Society for Psychical Research]. I find it very hard to account for the periodicity of the striations on the assumption that they are produced by reflection of a bright light from the lens, or projections somewhere within the camera body.'

He concluded of three photographs depicting the intrusion of a greenish-white sphere, 'These green objects, if real, are of such size and unusual form that they could hardly have escaped notice. I cannot understand how they could have been produced by stray light within the camera. The "green globes" are too clear and detailed for that.'

Mrs Smethurst was given some film and asked to try and induce some of the images. Later, Peter Hough visited her with a loaded camera and asked her to use up the film in his presence. No spurious images showed up on any of the films. Since then Mrs Smethurst has produced another picture which depicts the fuzzy black line – with yet another camera. In his report, Vernon Harrison observed 'The fact that these phenomena are being reported, not by professional magicians, tricksters and fake mediums, but by ordinary people using ordinary film in ordinary cameras, makes them all the more worthy of attention.'

Ted Serios

You could not get more 'ordinary' than Ted Serios, a chain-smoking Chicago bell-boy who liked a drink! Not only was he able to produce anomalous shadows and lights – known as 'blackies' and 'whities' – but the recognizable images of buildings.

Ted Serios produced dozens of 'thoughtographs' by concentrating his mind on pieces of film. He is seen here in action

Serios rose to fame in the 1960s. It was in April 1964 that Dr Jule Eisenbud, Associate Professor of Psychiatry at the University of Colorado Medical School in the USA, visited the man in order to carry out some experiments. Eisenbud was convinced he would uncover a fraud. In his first meeting with the 'thoughtographer', Serios pointed a polaroid camera at his face and pressed the shutter. Dr Eisenbud pulled the print from the back and was confounded to see, not a picture of a face, but the image of a building.

He continued to study Serios for several years and never found the fraud he had originally set out to expose. Since then, many investigators have tested Ted Serios, supplying their own film and cameras. Sometimes they have held the cameras themselves, pointed at the psychic but the results are the same with anomalies for which there appears to be no rational explanation.

Sceptics looked for fraud and believed they had found it, in the form of what Serios called his 'gismo'. In the early days he just stared at the camera to produce paranormal images, but later he introduced a small plastic cylinder, covered at both ends with cellophane, and a piece of blackened film. Occasionally he simply rolled up a piece of paper. This was held in front of the camera, in order, he claimed, to stop his fingers accidentally obscuring the lens.

Paranormal debunker, James Randi, a professional magician, believes there is a more sinister reason for the device. Randi suggested that the tube is used to hide a small lens and a microtransparency. If held in front of the camera and hidden inside the gismo, the image on the transparency would be projected onto the polaroid film. Afterwards the device could be palmed away. Photographs have been produced using such a device, but in the hundreds of demonstrations given by Ted Serios, not once has there been the suspicion he was using one.

Many psychics use gimmicks as a psychological focus for their alleged powers, so perhaps the gismo should be viewed in this light. Eisenbud and other researchers took steps to ensure that such a device could not possibly be hidden in the gismo. Serios wore short sleeved shirts, or sat topless for the demonstrations, and was only given the gismo when he was ready to produce an anomalous print. Afterwards, usually no more than fifteen seconds, it was taken from him and examined. Frequently, when the camera flash fired, the researchers could see through the gismo, and knew it was empty. Sometimes the researchers have held both the camera and the gismo, with paranormal results. All that seemed necessary was the presence of Serios. There are other reasons too why James Randi's explanation seems untenable.

American psychical researchers Dr J. G. Pratt and Dr Ian Stevenson who conducted approximately eight hundred trials with the thoughtographer, commented that the very nature of the images ruled out the device theory. Lettering on buildings produced on the prints was often misspelt. In tests, target pictures hidden in sealed envelopes were used. Serios has paranormally reproduced these images on polaroid film. Many of the photographs depict architecture which was subtly changed from the target buildings. These seem to be products of the past, future, or a distortion of reality.

A significant experimental session took place at the Denver Museum of Natural History on 27 May 1967. Surrounded by paleolithic and neolithic artifacts it was hoped Serios would produce images from prehistoric times. He began by concentrating on an impression of a primitive man lighting a fire. Several strange images were produced, but the best showed a Neanderthal man in a crouching position. Professor H. Marie Wormington of the Department of Anthropology, Colorado College, recognized it immediately as very similar to a life-size model which resided in the Chicago Field Museum of Natural History. Postcards depicting the full Neanderthal tableau were available for sale.

Was James Randi right after all? Had Ted Serios surreptitiously held a transparency in front of the camera lens? Detailed examination of the two images showed that they were not identical. The angles of posture of the Serios Neanderthal were subtly different from the exhibit. Further, several professional photographers and photogrammetric engineers, went on record as saying that the image could not have been produced from exposing the camera to a single microtransparency.

This was the dawning of the end of Ted Serios' enigmatic powers. They had waxed and waned before, but within a year all he could produce were blackies and whities. The bell-boy who had risen from obscurity sank back again without complaint.

The Veilleux brothers

Dr Eisenbud now turned his attentions to two brothers who were achieving some very remarkable results. Richard and Fred Veilleux of Maine, USA, began experimenting with an Ouija board in 1966. It was through this device that they were instructed to take photographs at certain designated places.

The first location was the grave of a little girl in the cemetery where the brothers worked as stonemasons. They took the picture on 1 August 1967. It showed the figure of the child standing next to her gravestone. She had been murdered some years before.

Photographs taken of relatives by the brothers and their wives, produced images which blotted the subjects out. Under the supervision of Eisenbud, one photograph showed two faces superimposed over two doors. These turned out to be a young and an old portrait of US Marshal, Jeff D. Milton. They were traced to a book called *Album of Gunfighters*, where the original pictures appeared side by side. As with Serios' picture of the Neanderthal man, it appeared the thoughtographers had been caught in a hoax. But similarly the images differed slightly from those in the book.

The brothers found some images superimposed over a photograph they had taken of their kitchen in July 1968. This was identified by

Professor Charles Lyle as an Australian rock painting. The painting appeared in Erich von Danekin's *Return To The Stars*, one year later, and was unknown before then. Lyle described the painting as a 'pagan Last Supper'. He was not to know that this aboriginal piece of art completely obliterated another. For hanging on the kitchen wall was a print of Leonardo's 'Last Supper'.

Other thoughtographers

Most anomalous images turn up on film spontaneously. Few have the power of control apparently demonstrated by Ted Serios. The spirit photographers of the nineteenth and early twentieth century were largely fraudulent – superimposing the faces of 'spirits' onto normal photographs for gain. One Spiritualist medium, however, who survived rigorous testing was William Hope of Crewe in Cheshire. He was approached by several investigators including Dr G. L. Johnson, a Fellow of the Royal Photographic Society. Dr Johnson published a book in 1928, called *The Great Problem: Does Man Survive?* in which he described the experimental procedures.

An assortment of photographic plates was purchased from several London suppliers and taken by Dr Johnson to Crewe. There, Hope was not allowed near the plates, nor to approach other members of the party. The medium was then photographed using a number of different cameras. Once processed, the plates depicted a variety of images including recognizable faces of people living and dead, written messages, flower-like shapes and clouds of 'ectoplasm' – a psychic substance out of which paranormal entities are said to materialize.

A further test was carried out with a sealed box of unused plates held to William Hope's head. When processed, the outer plates remained unexposed, but the two inner ones depicted more writing.

Not all thoughtographic images are of Earth-based people and objects. American, Stella Lansing, produced UFO type 'objects' on ciné film that rotated and flashed. Dr Berthold Eric Schwarz, a psychiatric consultant, investigated

William Hope was a controversial 'spirit' photographer who survived many tests, but was exposed by others. Here, the image of a man's dead son appeared on plates provided by the father who called on Hope unannounced

her case in the 1970s. All sorts of still and ciné film, including black and white 16mm infra-red were used in many different cameras, under the supervision of Dr Schwarz. As well as the disc-shaped UFOs, Mrs Lansing produced anomalous lighting effects, familiar but distorted landscapes, strange ghostly figures and faces – including a disfigured version of her own. Often filming was done on a 'hunch', on other occasions the results were entirely unexpected.

One of Stella Lansing's most fascinating images was obtained when she tried to film her favourite television programme. The result was a collage of objects superimposed over the show. But most impressive of all was the distorted image of a bearded, turbaned figure holding a flute. There is an expression of absolute misery on his face.

Prayer power

While researching this book, Peter Hough was sent a colour photograph together with a few details. He is not able to verify the picture at this stage but here is the story nevertheless.

The photographer, who is a woman with strong Christian beliefs, was travelling on a passenger aircraft between Australia and New Zealand when it flew into a violent electrical storm. The aircraft was being badly battered and appeared to be losing control, causing panic amongst the passengers. The woman began praying, then noticed another woman doing the same. She went over to her and they prayed together.

First of all they asked that the other passengers should be calmed and order restored – this was answered immediately. Then they asked if it was the Lord's will, that the aircraft should be delivered. After a while the storm began to abate and they were out of danger. The woman took a photograph of the receding storm through the cabin window. She noticed nothing odd at the time.

The aircraft landed safely. When the roll of film was eventually used up she took it to her local chemist for processing. She forgot about the film until the chemist wrote to her. In the letter he said the film was ready, but he was particularly intrigued by one of the photographs and hoped she could explain it to him. She was as amazed as anyone at the picture, which on the surface showed the storm clouds. Yet when rotated through 90° depicted a Christ-like figure.

There is no doubt that the woman saw the paranormal image as a sign from God that He had indeed heard her prayers and acted upon them to save the aircraft. Yet the evidence indicates that it is more likely that the holy image emanated from her own mind, rather than the hand of God.

Thoughtographic exposures

- There is firm evidence that anomalous images can appear on film which cannot be explained in normal photographic terms.
- It is easy to produce faked images on film, but proper test conditions can make that almost impossible.
- Thoughtographers of the order of Ted Serios and the Veilleux brothers are rare, but many people are probably producing 'blackies' and 'whities' without realizing they are paranormal.
- Many photographs of 'ghosts' which were not seen at the time, are probably thoughtographs and may have nothing to do with surviving spirits.
- The fact that some images, such as the Veilleux's aboriginal cave painting, could not have been in the mind of the thoughtographer, indicates that human consciousness acts as a channel for information from elsewhere.
- As in all areas of the supernatural, the 'Cosmic Joker' is not above invading the pack. Hence, perhaps, Serios' Neanderthal exhibit and the Veilleux's book illustration.

THE SPIRITUAL DIMENSION

RELIGIOUS MIRACLES

In Fatima, Portugal, crowds gathered during October 1917 as two girls communed with a 'radiant lady', identified as the Virgin Mary. On 13 October hundreds saw the 'sun' dancing across the sky

It has been said that faith can move mountains. While that particular feat might be hard to achieve, certainly faith, or belief, can often act as the catalyst for paranormal phenomena. Nothing is stronger in this respect than religious faith. It might not move mountains but it certainly seems to create the conditions for more minor miracles.

Visions of the Blessed Virgin Mary

Visions of the 'Blessed Virgin Mary' (BVM) have occured in most countries that have a strong religious culture, particularly those of the Roman Catholic faith. Most famous of these, concerned a fourteen-year-old French girl called Bernadette Sourbirous, who experienced visions of Mary at Lourdes. They started on 11 February 1858 and continued into July. It was on 25 February that the vision instructed her to dig in the earth. Three hundred people watched in amazement as the girl clawed in the ground and put earth into her mouth then vomited. Where she had been digging, a spring emerged which she washed in, producing 18 gallons of water a day. Although no one else had shared the vision, the spring became central to the world famous Lourdes' shrine, attracting thousands of pilgrims every year in search of miraculous cures.

Thirteen years later, in the French village of Pontmain, two young brothers claimed they could see a beautiful smiling lady looking down at them from a position in the sky above a house. Their parents and a neighbour could see nothing, but the visions persisted, and during the evening a crowd gathered, including some more children. Before they had a chance to talk to the brothers they were asked what they could see. All the children described the same phenomenon, and indeed, all of them reacted at the same moment when 'Mary' spoke to them. Just over three hours later, the figure became obscured by a mist and disappeared.

One of the best known cases took place at Fatima in Portugal, in 1917. There two little girls communed with a radiant lady who was identified as the Virgin. As the visions continued great crowds gathered, and although they could not share the vision, the crowd claimed they saw the sun dancing across the sky. Mary passed on some prophecies to the children which it is claimed the Catholic Church has suppressed.

In the modern world, visions of the BVM are quite common. It has been suggested that the visions take place prior to some major upheaval. Was it coincidence that one of the major cases which captured the attentions of the Western media, occurred in Yugoslavia in the 1980s, just a few years before the terrible civil war which has devastated that country's culture and people?

Moving statues of the Virgin Mary

The shrine at Lourdes was copied around the world. On 21 August 1879, Mary McLoughlin of Knock, Eire, reported seeing the Virgin Mary. Subsequently a shrine was built which has since attracted millions of pilgrims. It was on Monday evening, 22 July 1985, that two women, passing the grotto at Ballinspittle, County Cork, were convinced that they saw the life-sized statue of the Virgin move.

The statue itself is firmly set in concrete, standing before an artificial rock cave 18–20 feet (5–6 metres) above a dais, and weighing around 3 cwt (152 kg). Seventeen-year-old Clare O'Mahoney and her mother, Mrs Cathy O'Mahoney, claimed they had seen the statue rocking backwards and forwards. The women were joined by about forty other people, including a sceptic, who had gone to see what all the fuss was about. They claimed they had seen the statue move so dramatically at one stage, it was feared it might topple over and fall on some children.

The following night fresh witnesses described variously how the statue's head and shoulders moved, how it swayed from side to side, backwards and forwards, shivered, how hands and lips moved, and the head nodded. By Wednesday the crowds had grown to hundreds and the firmly embedded concrete figure continued to move. The numbers were swelled to thousands before the end of the week. Ironically, several confirmed Catholics came away having seen nothing, while sceptics reported seeing movement. One of these was *Sunday Tribune* journalist, Eileen Forristal, who on 4 August was shocked to realize that the statue was swaying from the shoulders upwards.

On 31 July, many viewers of a news bulletin concerning the statue, telephoned the television company (RTE) to report having seen the face of Christ superimpose itself on the screen. A rerun of the film showed nothing.

At dusk on Hallowe'en, three well-dressed men appeared at the grotto. They harangued those there for worshipping a statue and believing it could move, then set about it with an axe and a hammer. The men were caught in a police road block, and it transpired they belonged to a religious sect opposed to Mariolatry, the idolatrous worship of the Virgin Mary. Despite this, worshippers continued to see the damaged statue move.

In the wake of the Ballinspittle phenomenon, there were reports of other moving statues elsewhere in Ireland. A week after the initial events at Ballinspittle, witnesses at nearby Courtmacsherry noticed their own icon moving alarmingly, although a priest who held onto it felt only a slight tingling. A statue which had just been erected at Mount Melleray, County Waterford, was reported by several children to have left its pedestal and walked down towards them, passing on a message of the dire state of the world. Although the children were sincere, others at the shrine saw nor heard anything.

There were many other reports all over Ireland, some incidents taking place in churches. It also emerged that the statue at Dunmanway had been on the move for the previous five years. There were reports too of demonic visions at Ennis Cathedral and in the church at Mitchelstown, County Cork.

Moving explanations

Church officials in Ireland were careful at this time to retain a position of objectivity. However, some clergy claimed to have shared the experiences, and others, such as Dr Thomas

McDonnell, Bishop of Killala, were openly sceptical.

The world in general, and the Irish Republic in particular, were in the grip of terrible natural, man-made and economic disasters at the time of the Marian phenomena. This caused many sceptics to cite the experiences as a psychological response by those seeking hope and supernatural confirmation in God. There was also discussion in some of the movements being due to hoaxes, although no one came forward with any evidence.

Dr Kirakowski and four colleagues from the Psychology Department of University College, Cork, visited Ballinspittle and came up with two theories. One was that heat generated by small light bulbs in a halo suspended above the statue might cause the air to shimmer, giving the illusion of movement – although this could not apply during adverse weather conditions. The other was a physiological reaction to staring at the statue for long periods which would cause the muscles in the back of the neck to shake, until the image appeared to move. This can also happen in the case of a bright light in a darkened room. An 'autokinetic effect' causes the brain to perceive the light moving, when in fact it is stationary. While such theories might provide a partial explanation, they cannot apply to movements noticed during daylight. Were the phenomena which followed Ballinspittle, due to progressive hysteria as some suggest? How then could sceptics be drawn into this, many whom lacked religious belief?

Religious images that bleed and cry

If plaster statues of the Blessed Virgin can move, is it conceivable that icons of Mary and Jesus Christ can weep and bleed? There is much documentary evidence stretching back hundreds of years that this is the case. In 1527 a statue of Christ was said to have wept copiously. Dr Piero Casoli studied weeping Madonnas in the 1950s and said there were at least two cases every year in Italy alone. One of these occurred at Syracuse in Sicily, in August 1953. The Madonna was an ornament presented to Mr and Mrs Janusso as a wedding present. Antonietta Janusso began to suffer from mysterious illnesses. One day she looked up at the figure from her bed and noticed it had begun to weep. The phenomenon attracted many witnesses of high repute over the next few days. Suddenly Antonietta began to feel better and by the time the statuette stopped crying, she was recovered.

There are numerous photographs showing what looks like blood issuing from the palms, head and legs of statues. These correspond with the wounds of Christ, except we now know that during crucifixion the nails were driven through the wrists, *not* the palms. Could fraud explain the phenomenon and is the red substance really blood?

Trickery does not seem to be an option in most cases. In April 1975, a plaster statue of Christ began to bleed from the hands at the home of Mrs Anne Poore, in Boothwyn, Pennsylvania, USA and repeatedly did so every Friday and on holy days. Eventually it was moved to St Luke's Episcopalian Church. Father Chester Olszewski confirmed that fraud was not involved. He had removed the hands and found them to be solid wood, yet they continued to bleed for up to four hours. X-rays confirmed the statue contained no hidden reservoirs. The red liquid was analysed by Dr Joseph Rovito, who confirmed that it was indeed human blood yet the results were contradictory. There were hardly any red cells in the blood indicating blood of great age. However, the blood always flowed some distance before coagulating. This was the behaviour of fresh blood which would contain millions of red blood cells.

The phenomenon of weeping and bleeding statues has been compared to poltergeist activity. Peter Hough investigated a case during 1989 which draws a comparison. The case concerned a woman living alone who was plagued by various inexplicable activities. One of these was the materialization of water in the house. The woman discovered a book wet like a sponge, water dribbling from a radio, several inches of water in a metal bin, and a large quantity which soaked her mattress, when only seconds before there had been no water whatsoever. One can see how the phenomenon could be applied to the statues.

A bloody miracle

Can the dried blood of a saint suddenly turn to liquid through supernatural intervention? Tradition has it that St Januarius was the 13th Bishop of Benevento in Italy. He was beheaded by the Emperor Diocletian around AD 305 after he had been thrown to the lions who refused to attack him. Several relics associated with him were preserved, including remains dug up a century after his death. As the corpse was being

A typical example of the many religious icons which have been seen to 'weep', this statue of Our Lady is situated in Maasmechelen, Belgium. Exhaustive scientific tests seem to prove that this is not brought about through trickery

transported to Naples, it allegedly began to bleed. A woman called Eusebia collected some of the blood in two phials which were sealed afterwards with a putty-like substance. This blood spontaneously liquefies. Currently, this happens about eighteen times a year during mass, and it has been photographed and filmed. The circumstances of the 'miracle' are many and varied. It has been recorded in detail since the fifteenth century. There is a huge body of testimony, contributed to by many intelligent, sceptical and scientific persons. No one can doubt that it does take place. Although the process has not been observed under strict laboratory conditions, in 1902 spectroscopic analysis at the University of Naples confirmed that the substance was indeed blood, adulterated with other unidentified materials. This quashes some assertions that it is purely a chemical concoction.

Proponents of that theory reason that the liquefaction takes place at a certain temperature and humidity. However, temperatures have been recorded as low as 15°F (–9°C), and as high as 86°F (30°C). The blood has taken as little as ten minutes to liquefy and as long as twenty-four hours. The length of time does not seem to depend on the ambient temperature of the room. When the material changes, it is a sudden process that turns clotted blood into a frothing liquid. Modern sceptics have tried to duplicate the process but with incomparable results. Naturally, perhaps, the Church is opposed to the opening of the phial, which would result in the destruction of the ancient relic.

Stigmatism

Can followers of Christ replicate the bloody wounds of His crucifixion? There are hundreds of documented cases where blood has spontaneously seeped from the head, hands and feet – open wounds appear and whiplash marks score across the body. Such people are called 'stigmatics'. Once again, the earliest known accounts date back to medieval times. It was in 1224 that St Francis of Assisi became stigmatized.

Thomas Celano, writing two years later, maintained that St Francis had just had an encounter with an angel when the marks appeared. His hands and feet bore the appearance of nails through them, and a scar materialized in his side as if caused by a lance. This latter often bled, staining his clothes.

The stigmatic Padre Pio displaying the wounds on his hands which resemble the wounds of Christ on the cross

Today there is undeniable proof preserved on film of stigmatics spontaneously bleeding, usually on religious days. In 1901 Gemma Galgani, an Italian girl suffering from spinal tuberculosis, experienced the agony of crucifixion while praying one Friday in March. She was discovered covered in whip marks and blood. During the next two years until she died, the stigmata regularly appeared. On occasions,witnesses saw deep wounds appear through her palms, then, later, closing just as rapidly, leaving a white mark.

Although most religious miracles are firmly connected with the Roman Catholic faith, this is not always the case. Cloretta Robertson, a black non-Catholic ten-year-old from Oakland in California, started bleeding from the palms nineteen days before Easter in 1972. Blood issued through her skin for a few minutes every day. When Good Friday arrived, the phenomenon stopped – never to return.

Examination of the blood from stigmatics has found it to be clean and disease free. Some wounds remain open for several years, defying natural healing processes, and failing to become infected. Is stigmata an example of 'mind over matter', an illustration of the power of the mind over its physical host? Experiments where stigmatics have been hypnotized in attempts to induce the phenomena have been unsuccessful. Yet there is evidence that the wounds stem from imagery stored in the mind. As previously indicated, crucifixion was through the wrists, not the palms – where the skin would tear, unable to support the weight of the body. Yet stigmatics bleed from the palms, exactly as the crucifixion is portrayed in works of art.

Cured by love

The idea that stigmata are an historical phenomenon associated with constantly devout nuns and priests took quite a battering during 1993 when a woman from Lincoln, England, told her remarkable story.

Heather Woods, a 43-year-old widow, claims not only to have the wounds of Christ as absolute proof of her faith but to have been cured of life-threatening cancer as a result.

The first that she knew of this miraculous event was after being diagnosed as terminally ill with virulent cancer in 1990. She had major surgery which failed to remove the disease and spent time in a hospice awaiting her end. Throughout her ordeal her devotion to God remained strong.

Then, in May 1992, she saw a vivid image of Jesus on the cross and felt herself being drawn towards Him. She was projected into an out-of-body vision where she was suddenly free of all pain and anxiety. Upon returning to normal consciousness, she noticed marks forming on

Cloretta Robertson, a ten-year-old non-Catholic, started to bleed from the palms days before Easter 1972. Stigmatics portray the wounds as popularized in art. In fact the wounds in crucifixion are through the wrists

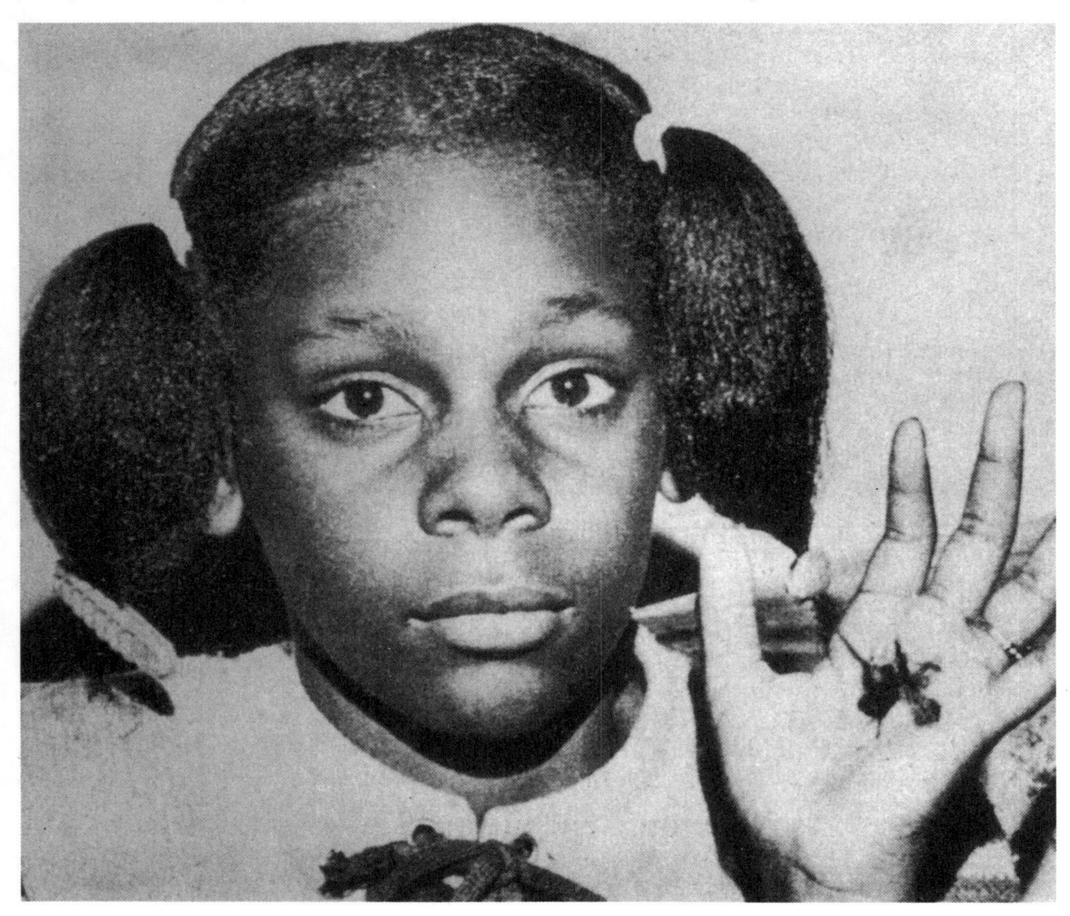

Miracle conclusions

- The complexity of religious visions by people of little education seems to rule out a psychological explanation as the entire answer. However, entities 'recognized' as the Virgin Mary seem then to take that belief onboard and act accordingly.
- Once miracle visions have been established, even some sceptics have shared in them.
- Scientific analyses of statues which bleed have ruled out trickery and have established that the blood is of human origin.
- There is no 'rational' explanation for stigmatics – people who spontaneously display the wounds of Christ. Stigmatics have been extensively examined.
- Religious belief systems seem to channel and shape supernatural phenomena so that they reinforce those beliefs. Many religious miracles can be compared with poltergeist phenomena. At root we have a common phenomenon shaped by an interaction with human consciousness.

her hands, feet and sides. At first she assumed these were blisters and as they appeared they felt like ants crawling over her body.

The physical effects lasted for three weeks, oozing blood for several hours each day. The mark of a cross was etched into her forehead and a halo or ring of thorns appeared on her foot. Her doctor was called and was stunned when he saw Heather. She told him: 'I am healed.' Indeed, it was to be discovered the cancer had faded away.

The stigmata reappeared at Easter in 1993 as predicted in a vision and were filmed by a television crew working on a religious affairs programme. Heather insisted that she had been chosen to be cured and to have this visible sign to prove to the world that miracles happen and faith is still important.

Heather's life has undergone a fundamental change because of this. At the same time as the stigmata she started to pour out pages of script by automatic writing, done in unfamiliar style with her left hand (although she herself is right-handed). These include detailed philosophy about love and skilled drawings, for example of Christ being crucified.

Whatever the case, the stigmata marks cannot be denied. They were there for millions of viewers to see.

APPARITIONS

A chance of a ghost

The idea that there are ghosts haunting the nether regions of our world is the oldest of all spiritual beliefs. Virtually every religion accepts that the human body has a less physical counterpart and that after death it separates from our mortal form and makes its way towards the afterlife. Equally, most such religions have some concept of these so-called 'shades' sometimes becoming earth-bound and roaming around familiar surroundings or haunts to be occasionally seen by terrified witnesses.

Most tribal cultures use a shaman as a conduit between their own world and the land where they believe their ancestors repose. The shaman can bring teachings from the dead that can assist the living. The Egyptians had a *Book of the Dead* which charted the soul's passage to the hereafter. It was buried with the body, along with various treasures, sometimes even pets of the deceased, to help ensure that they made the arduous passage.

In the Old Testament, prophets tell of visits by 'apparitions'. Indeed, as scholars have noted, without supernatural visitations and what might be interpreted as extrasensory perception (ESP) the Bible would be a thinner book.

However, just because a subject has universal acceptance among the many and varied cultures of the human race this does not make that belief a fact. Psychologists tend to argue that it is so commonplace simply because the mind's self-consciousness makes us uniquely aware of our own mortality, something none of us are capable of imagining. Individually we cannot conceive of ceasing to exist and the same is true collectively for our species. We simply have to assume that we continue after physical extinction.

The idea of ghosts may have become such a common myth merely to protect ourselves, as a very effective form of wish fulfillment fantasy.

Sensory phantoms

When asked to describe what they think a ghost may be, people have very specific ideas in mind. These are clearly based upon an overdose of Victorian novels, TV series and low-budget movies. As such, the ghost is assumed to be a floating ethereal form, reflected at Hallowe'en by children wearing white sheets over their heads, or otherwise a semi-transparent spectre that you can see through and walks through walls.

Yet the reality of what an apparition looks like is very different according to witnesses' accounts. The most common problem related by a ghostly percipient seems to be how to recognize that they have not just seen an ordinary person. The ghost generally looks and acts the same as anybody else and is often not known to be a phantom until well into the experience. Then either the figure inexplicably disappears, or else information is discovered to indicate that the seemingly normal person was at that moment already dead.

Ghosts, in other words, are pretty normal and suffer only from the handicap that they are not supposed to be alive. Indeed, most witnesses point out that it is more than possible we could all see ghosts on a regular basis without realizing that we had done so. They would seem

The ghost of a monk gliding down the steps beside a church altar. Sceptics point to the elongated facial features that might suggest a mask. Nothing was seen at the time by the photographer

Some apparitions come in the form of an amorphous shape. Children often seem more able to experience them than adults. In 1991 this 2-year-old began to say 'Old Nanna's here' and point up in the air. Some researchers believe this photograph may show him seeing the ghost of his great-grandmother

so ordinary that under most circumstances we would be unlikely to tell the difference between a ghost and someone just walking down the street.

Equally, although the view is that ghosts are visible forms – apparitions – all senses are affected. In the cellar of an old jeweller's workshop in the Roman city of Chester several of the staff told Jenny Randles about 'George'. Nobody has ever seen him, but he has often been heard, footsteps echoing on a stone staircase. Research after the fact showed that the building, near the city's North Gate, was built atop an old jail where condemned prisoners were kept three centuries before, awaiting the hangman's noose. The belief seems to be that the footsteps are auditory replays transmitted across time.

Ghostly smells are common too. Actress Doreen Sloane tells of the ghost that lived happily in her flat in Birkenhead, seen by all family members but most typified by the phantom cooking smells that invaded the building. And no fewer than one in three recent widows or widowers report smelling their spouse's perfume around the house in the weeks following their death.

Sounds are also common, with various well-attested cases of noises from long past battles being heard by visitors to an area who were sometimes not aware that any such battle had ever occurred there.

Research by Celia Green of the paraphysical institute in Oxford found that there are ghosts of touch, even taste and emotion, as well as the more common smell, sound and vision experiences. Visual apparitions were particularly common and differed markedly from psychopathological hallucinations, that is those experienced by sufferers from mental illness or whilst under the influence of drugs. Ghostly apparitions were far more mundane with less exotic features ascribed to the hallucinations.

A typical spontaneous case, one of tens of thousands now documented the world over, comes from Elizabeth Dempster of London. As soon as she moved into an old building in Highgate she picked up its sombre mood. She tried to fight it by repainting the walls in bright shades, but this did not help. Then, one night, she awoke with a sense of a presence in the room. At the foot of her bed she clearly saw a woman in old-fashioned dress wearing her hair in a bun. She could take in quite specific details and despite the dark see the sad, hopeless expression on her face. Then the figure simply disappeared.

When researching the history of the house, Elizabeth traced the mystery woman from a time soon after it was built. Her description matched an Italian woman whose husband had died in tragic circumstances. So desperate was she after this loss that she had locked herself away in the bedroom until she herself had passed away. It was as if that terrible concentrated emotion had become bottled up and retained by the building across the years. The emotion itself was perhaps the easiest to tap into, as Elizabeth Dempster did the moment she moved in, but for that one night the conditions were somehow right to allow the new occupant of the house to gain a brief view of its former tenant.

All in the mind?

The most common type of apparition is what is called the bedroom visitor, exactly like the one in the old Highgate building. This immediately suggests a possible solution. For if ghosts tend to occur as we wake from sleep, how can we be sure that we really do awake? Might they not simply be dreams or visionary experiences which we wrongly interpret as real events?

Hypnagogic experiences are well known to psychologists. They occur in about two-thirds of people as they slip from waking consciousness into sleep. Common among these so called hallucinations are images of faces. They are transient things usually swallowed up by a period of unconsciousness so few of us retain any memory of the phenomenon unless special efforts are made by psychologists to study this imagery. They do this by waking the subject as soon as their brain wave patterns suggest they are seeing an image and asking what it was that they visualized.

It is thought possible that some 'bedroom-visitor' ghosts occur when a frightening image is so vivid that it propels the person into sudden wakefulness and they then retain memory of what was an hallucination. They then falsely, but understandably, assume that what they saw was a real encounter with an apparition.

However, while this idea may well work in some cases it struggles to explain those where the vision seems to be confirmed by other evidence. How could a hypnagogic hallucination be of a former occupant of a house that the person undergoing the hallucination had never

Animal ghosts are often reported, such as this 1926 photograph of a phantom dog appearing at the rear of an Irish wolfhound. It was identified as a cairn puppy which had died 6 weeks earlier, and had been an inseparable friend of the Irish wolfhound

seen or heard about? Certainly, it seems likely that all apparition or 'bedroom-visitor' experiences happen while the witness is in an altered state of consciousness and may even be completely visionary in nature. But this by itself does not account for the source of the image in the first place.

Studies have also been made into other forms of recurrent imagery, associated with schizophrenia. There is strong evidence that apparitional experiences seen in such circumstances are similar to those witnessed by people who are not schizophrenic and merely have a one-off ghostly encounter. It seems unlikely that ghostly witnesses are closet schizophrenics. More probably both share some process in the brain.

In March 1979 a teenage girl had a typical 'bedroom-visitor' experience of a small figure with penetrating eyes that entered her bedroom in Shipley, West Yorkshire. She heard the floorboards creak as the figure approached, then it simply vanished.

The interesting difference here is that Nina was an epileptic and she was in the middle of a seizure at the time. Epilepsy is somewhat like an electrical storm within the brain and it is known to lead to occasional hallucinations. Did Nina somehow synthesize the same changes within the neurones of the brain which in other circumstances and with non-epileptic witnesses also leads to an apparitional experience?

Dr Wilder Penfield, a neurologist, found something odd during brain surgery, which can be carried out while the patient is conscious as no pain is experienced. If a certain neural pathway was accidentally touched they could suddenly experience an all senses action replay of a fleeting image. Frequently the patient was unable to locate this as anything specific. It seemed very odd, almost visionary.

Presumably it was one of the trillions of memory bits believed to be stored in the huge network of electro-chemical circuits that make up the brain, most of which are not accessible to our conscious recall. If the same process can occur during an apparitional experience then perhaps the ghost is sometimes nothing but a stimulated vision created out of a momentary short circuit taking place within the brain.

Ghosts that are not dead

Most paranormal researchers reject these reductionist approaches and believe that something more than hallucinations must be involved. This tends to suggest that we expect ghosts to be what people have thought them to be for many centuries, that is the souls of the dead trapped on earth.

Of course, the real problem here is that there is no hard evidence that a soul exists. Indeed there remains a big debate about whether mind is a sort of energy field that merely interacts with the mechanical brain or whether the mind is no more than an epiphenomenon created in the brain and having no possible life after the brain dies.

Those who believe that brain and mind are one thing have no recourse to a belief in ghosts. They cite an analogy of a car. Whether one car goes fast or slow, runs well or poorly, depends upon its components. Destroy its engine, the vehicle's brain, and the car will be dead. It will no longer have abstract concepts like speed or performance, because these were just side effects of the mechanical prowess of the vehicle which itself totally depended upon its physical properties.

On the other hand, those psychologists who think that one cannot reduce all mental activity to functions of the brain say that there must be a mind working in tandem but apart from the brain. They find it easy to accept that some aspect of this could trigger ghostly experiences.

These researchers cite the analogy of a radio set. This plays all manner of things from music to speech and seems at first sight to do so because of its components. Smash the radio to bits, or even damage one of its key components beyond repair, and things will go badly wrong. The radio may work less well or in extreme cases stop working altogether and never function again. However, after this destruction all that it was 'producing' is still there,because these are really unseen radio waves which were merely being brought into our perception by the circuits of the machine. The brain may similarly harness the energy of the mind and channel it into the reality of the physical world.

This dualist view has evidential support. For instance, many psychological studies suggest that mind is superior to matter (hence the phrase mind over matter). There are well-attested cases where thinking something strongly enough creates measurable physical changes in the body. That is why a placebo works – a placebo being a useless pill given to a patient in order to alleviate illness. The patient thinks it is medically effective and thus helps ensure that it becomes effective, even though there is no medical case for this. If mind were an epiphenomenon of a mechanical brain it would be hard to explain this superiority.

Similarly, most paranormal phenomena involving consciousness, for example, precognition, ESP, etc., only seem to work if we conceive of mind as a timeless, spaceless energy field capable of extending beyond the limitations of the brain and touching other energy fields in the universe. Non-dualist researchers must argue that concepts such as ESP are therefore nothing but a myth.

However, those who claim that mind and brain are one thing and so ghosts cannot be souls of dead spirits have unexpected support from some of the evidence. Contrary to all expectations, you do not have to be dead to be a ghost.

There are a number of well-documented cases of what are called crisis apparitions. Usually a witness sees a friend on the verge of death and the typical interpretation is that the friend has made one last stop on the way to heaven, perhaps to say goodbye. In one case like this a woman from Cheshire received a telephone call from her husband, who reported, 'I am very far away', at the exact moment, as she later discovered, that he was dying after being shot by a car thief. The interesting twist here was that the woman had no telephone. She was doing the ironing, heard a phone ring, picked it up without thinking and was then struck by the strangeness of her husband's words. As his voice faded away she realized that the experience was an hallucination and she was clutching thin air.

However, in quite a few other cases the person was not dead at the time, nor indeed were they even close to death. In one report a man saw the phantom of a friend walking by his house in such a way that mistaken identity was highly unlikely. Yet at the time she was alive and well at home but was suffering a nervous breakdown.

Clearly in instances such as these we need another explanation. They must be hallucinations rather than trapped souls. Presumably they are occurring in the mind of the witness who is somehow visualizing them as real events, perhaps after picking up the distress of the other person. This is implied by the way in which ghosts often wear clothes with which the witness was once familiar. In one case the ghost was seen as possessing only one arm, which was amputated just before death in a vain attempt to save his life.

What do we assume here? Perhaps that in the afterlife a soul is rendered armless to match the body? If so where do we draw the line – is someone blown to bits in a battle a disembodied bag of bits? The absurdity is obvious. Or was the ghost perceived in this way because the image was just a product of the mind of the witness and this was an appropriate image that represented this dead individual in their recent memory?

Do-it-yourself ghosts

Sceptical researcher Frank Smythe told how he invented a ghost. It was quite deliberate. The idea was to come up with a story and legend of a bogus spook, spread it around and see what would happen. He reports that several witnesses started to claim they had seen the phantom in the London Dock area where he had chosen to locate it. When he told them it was pure invention one refused to believe him, implying that he must have thought that he made it up but had tuned in to a real haunting.

A similar experiment was carried out with more sophistication in Canada, where a team of researchers pooled their resources and invented 'Philip' a long-dead English nobleman, giving him a life profile in as much detail as would anyone writing a historical novel. Then they set about holding seances for month after month, treating 'Philip' as a real person and trying to make contact with him. As a result various strange phenomena reportedly occurred and these contacts with the spirit of the made-up ghost became self-generating, taking on a vivid life of their own. After a year they had difficulty controlling the experiment because 'Philip' had become so real that he was behaving like a real ghost. As the researchers concluded, despite being a product of their minds something had animated this phantom.

Dr Morton Schatzman, a psychiatrist, had an equally extraordinary case of a woman called Ruth. She was American but lived in London. Ruth was experiencing frequent apparitions. However, they were not ghosts of dead people, but most often visions of her father, who was alive and perfectly well thousands of miles away across the Atlantic.

Instead of trying to stop her having such hallucinations as most doctors might, Schatzman helped Ruth to control them After some effort she could create lifelike, three-dimensional, real apparitions so well that she could practically do it to order. She could even produce a phantom Dr Schatzman whenever she needed him.

Experiments were carried out with Ruth's apparitions. She could make them turn on a light in a darkened room and see the room now as if it were lighted. But if she was then given a book and asked to read it, she could not do so.

Dr Peter Fenwick, a specialist in brain waves, worked with Ruth and she was wired up to equipment measuring brain activity. A pattern of images was placed in front of her retina,

The traditional form of an apparition as displayed in fiction – the white floating shape – here taken at the haunted Raynham Hall in Norfolk in 1936. The misty image is thought by some to be the famous 'Brown Lady' often seen in the building. However, to what extent does expectation cause us to read structure into an amorphous mass?

Authorities

- *Hilary Evans* is an excellent modern authority who writes lucidly and with perception about the nature of these phenomena. His work is best appreciated via companion volumes *Visions, Apparitions, Alien Visitors* and *Gods, Spirits, Cosmic Guardians* (Aquarian, Northampton, 1984 and 1986)
- *Celia Green* and *Charles McCreery* work for the paraphysical institute in Oxford. Their scientific research offers the best statistical analysis of apparitions, dreams and altered states of consciousness that is available. The most relevant work among a wide range of valuable studies is *Apparitions* (Hamish Hamilton, London, 1975)
- *Andrew MacKenzie* is one of the best 'ghost hunters' amid a field where stories often take precedence over research. His investigations into spontaneous apparitions are well documented. A good reference book is his *The Seen and the Unseen* (Weidenfeld and Nicolson, London, 1987)
- *Iris Owen* and *Margaret Sparrow* relate the background and results to their intriguing experiment to create a ghost in the invaluable *Conjuring up Philip* (Harper and Row, New York, 1976)
- *Dr Morton Schatzman* gives a full report on his extraordinary research with Ruth and her quite remarkable hallucinations in *The Story of Ruth* (Penguin, London, 1981)

designed to trigger specific responses that fed from the eyes to the brain. Ruth's eyes and brain behaved quite normally. She was then asked to create her apparition and place 'him' in front of the images, thus seeming to block their path to her retina.

Ruth did as asked, but of course, although she believed that she could see the apparition, there was nobody present in the room. As a consequence the light signals should still have got through to her retina travelling straight through the non-existent figure that she was merely imagining to be present. The brain waves would behave normally and prove that she was still seeing the flashing images. That was the theory, but the fact was different. When she placed the apparition that was not really there in front of the flashing images, her retina and brain waves responded just as if a real person had stood there to block out the path.

In other words, Ruth's apparition was clearly not real, at least in the sense that you or I could have seen it had we been there in the room. But it was real enough to superimpose itself upon the functions of her brain. The image was, if you like, undeniably an hallucination and a product of this woman's mind. But that mind had clear control over the workings of her brain to the extent that the apparition had more reality than we might care to expect.

Action replays

The manner in which many ghostly experiences seem to be action replays is considered a possible method by which science might explain them, without the need to resort to a concept of life after death.

It has been found that the most common denominator among the creation of apparitions is stress and emotion – for example, as we have seen in prisoners waiting to be hanged, a woman who pined to death after her lover's death, or the slaughter of the battlefield. This strongly suggests that certain events can be imprinted into the fabric of time and remain there.

In this regard, they would be a freak circumstance, similar to fossils in a rock formation freezing a moment in history from millions of years ago. We have fossils because a very unusual combination of events must have occurred at the moment an animal died to ensure that its bones, or more commonly, a cast of them, was permanently preserved.

If emotion or consciousness is an energy field then it is possible to imagine a way in which a particularly traumatic or sudden death may create a ripple in the environmental energy field of the earth. An analogy would be dropping a stone into a pond and seeing ripples spread out. Here the kinetic energy of the fall is transferred into motion of the surface water. If you imagine one of these waves rippling towards the shoreline where it then causes a waterwheel to turn, that movement by the waterwheel would happen seconds or even minutes after the stone was dropped, indeed long after the person who had dropped it may have disap-

Are some ghosts video replays of a past scene? This photograph at the University of Missouri in 1967 may show 'Myra' – an apparition frequently witnessed in the grounds. This could be a scene from 100 years before constantly replayed like a looped tape

peared. Indeed the stone dropper could die before the waterwheel turned, but that would have no effect on the turning of the wheel. The crucial event was the dropping of the stone which set in motion a chain of energy that outlasted its brief moment in time.

As such we can see how it might be possible that a person visiting a 'haunted house' could detect a ripple in the energies surrounding that building which were set in motion long before by someone now deceased.

There do seem to be some clues emerging that support this theory. It has been found that certain types of rock or stone from which buildings are created are better suited to ghostly replays. Equally, when there is building work or other activity on the land such as ground excavations, this can trigger a latent signal into action.

Perhaps critical is the presence of some type of crystal. Quartz is being actively researched in this regard, as found, for example, in areas or buildings where sandstone dominates. Possibly this somehow amplifies a latent signal rippling in the surrounding energy fields.

However, there also seems to be a further component and that is the person who visits the house or lives in the location. It may well be that there has to be a resonance between the electrical energy patterns within their brain and the location in question for the latent signal to be detected.

The theory presently suggests that there could be three important factors in the occurrence of an apparition. Firstly, a latent signal has to be present in a location, possibly etched as unseen ripples in the fabric of space-time at a building or site. Secondly, the location has to be visited by a person who is able to accidentally tune into that signal. Possibly they may have a resonance in brain-wave patterns and perhaps entering an altered state of consciousness brings this about. It seems to be a relatively rare, presumably chance, event not happening even with susceptible visitors all the time.

Phantom facts

■ Almost every culture on Earth has had a concept of the survival of the human spirit after bodily death. These all allow for the possibility that this spirit (or 'shade') can return from the afterlife to 'haunt' our material reality if certain conditions are not fulfilled. Most religions develop around this basic tenet.

■ Reports of ghostly apparitions have persisted throughout the history of civilization. They are found in the Bible, in texts of Greek and Roman scholars and throughout all more contemporary cultures.

■ Today psychologists argue that many, if not all, such cases can be accounted for by experience within altered states of consciousness. We are only slowly recognizing what these are. They include hypnagogic imagery (powerful visions on the border between sleep and wakefulness), false awakenings (where the person thinks they are awake and observe reality but are, in fact, still dreaming) and lucid dream states (where the dream is so vivid it appears like real life and conscious thought can for a time intervene and control what happens).

■ The accepted view of psychology is that mind is an abstract concept fully dependent upon the brain. It can have no separate existence and so cannot survive bodily extinction. An alternative theory sees the mind as an energy field that operates through the brain, just as radio waves are decoded by a radio set and can exist even if that set is destroyed and those radio wave energies can no longer be decoded. The fact that apparitions of living people can be seen suggests the possibility that there is an energy field that is amenable to detection.

■ A much-researched theory argues that apparitions are like echoes of emotional events imprinted as an energy field and overlaid onto the natural field of a place or building. As such the ghost can be 'decoded' by someone's brain who is passing by, or activated like a video replay.

Thirdly, some external factors may boost the signal and increase the likelihood of harmony between the visitor and the latent recording. For instance, building work might create vibrations in the microscopic crystals which make them more active than normal. It may also be found that ionization in the atmosphere is a trigger that boosts the signal. Either way, if this theory is verified, a ghostly encounter is the result of the right person being in the right state of consciousness at the right moment of time while being at a location where there is already a latent recording waiting to be tapped into.

As the action replay seems to decode via the senses of the witness, they may only experience a mood, presence, feeling or sound, according to how adept they are at translating material picked up by their subconscious mind. There is evidence from psychological studies that witnesses to apparitions appear unusually gifted with visual imagery. This could be a vital clue. In this rare combination of circumstance, they may 'see' the event replayed before them just like a video tape. Of course, this would not be a real event in the same sense as a video recorder might replay a tape. It would take the form of a very lucid hallucination in the perception of the person who has 'tuned in'.

This might well explain why photographs of ghosts are always disappointing. There are too few of them and the experts agree that most do not provide any evidence for surviving spirits. It would also explain why most apparitions involve the figure repeating an event from life and displaying little, if any, evidence that they are aware of the person who witnesses them. They do not do so, in exactly the same way as Humphrey Bogart or Greta Garbo may appear before us on a cinema screen and act out realistic emotions filmed years ago but which can obviously have no direct awareness of our presence.

However, this solution, if proven, might suggest a physical basis for apparitions, rooted in hard science. Equally it opens up the way for exciting future technology. Perhaps one day there will be a machine that can tune into these ghostly ripples. Then, instead of watching the latest soap opera, we will have the ability to decode real historical scenes and watch events from long ago brought back into our perception by science. It could be that in the future we will all be sitting down and watching ghosts on our television sets.

THE NEAR-DEATH EXPERIENCE

Dead men do tell tales

> *. . . I was first hit right between my thumb and the rest of my hand. By the time I shook that off and I was about to start back up, the rocket went off and exploded . . . I can recall doing a somersault . . . That was about it until it appeared to be about a couple of hours later . . . I could see the guy that pulled my boots off . . . it was just like I was looking right down on it . . . I could see me . . . it was quite obvious I was out of it . . . I looked dead. They put me in a bag.*

This is not a scene from a nightmare or a chilling novel. It is told as the truth. The narrator is a former soldier and he was talking to American cardiologist Dr Michael Sabom. What he describes is one moment in June 1966 – a moment when he died in action amid the terror of the Vietnam War. The young man was dreadfully injured with one arm blown completely away. So far as everyone surrounding him that day was concerned another life had tragically ended. He joined the list of statistics that shocked the world – just one more youth to be zipped into a body bag for the final flight home.

But this soldier was luckier than many. He did not really die. As his frightening testimony reveals he was somehow, somewhere 'conscious' of the futile efforts to save his life, of the pilfering of clothing and all the degrading rituals a battlefield can bring. This man screamed out in silent agony, desperately trying to make people understand. But no one heard him. Instead they took his body to the temporary morgue and began to cut into his veins and inject embalming fluid into them. The soldier awoke just in time to stop the mortician. This most assuredly was a near-death experience (NDE).

Few of us can imagine what it must be like to be alive and yet quite unable to convey that fact as these grim preparations continued; to be conscious, yet to be treated as a dead body. This is a disturbingly common situation. Modern medical advances have allowed many patients to return to life who only twenty years ago would have been deemed dead. Their experiences have become an increasingly reported phenomenon. Today this is one of the greatest unsolved riddles of science.

Oddly, virtually everyone says that they felt neither fear nor pain during the trauma. Most people found being dead quite enjoyable. Some did not even want to be revived and doctors tell of being cursed by their patients for saving pain-racked bodies and hauling them from the peace and serenity of this strange netherworld. But *is* that netherworld a reality – is it scientific evidence for life after death? Or, as some sceptics contend, is it merely an illusion brought about by the desperate death throes of a frightened (yet very much living) brain? Therein may lie the answer to one of the last true mysteries in the universe.

A modern history of death

Serious attention was first given to the mystery in 1969 with publication of a book called *On Death and Dying.* Its author was distinguished Swiss-born Elisabeth Kubler-Ross, who is an American-practising psychiatrist, widely considered the world's leading authority on death.

Kubler-Ross had counselled many through their final hours and had made the first real study of the psychological and physiological processes that attended them. She was convinced that at the point where life ran out, something very strange seemed to occur. Patients became happy, even joyous, sometimes spoke of seeing lights and were often more prepared for the end than was the medical team, even when it was unexpected. Her preliminary findings were a record of the peace and tranquility of those she nursed through this time and who, almost always, found it a pleasant experience.

It was not until six years later, with the publication of American psychologist Raymond Moody's work *Life After Life* that the near-death experience drew a wider audience. However Moody simply compiled a series of anecdotal cases of limited scientific value. But that list was to grab the attention of the world. It struck a chord in millions of people who seemed to somehow sense the importance of this research. The book soon became a best-seller around the world.

It was then recognized there had been centuries of reports of the same experience. Archives of psychical research societies, literary magazines, philosophical treatises and many other sources had mentioned what proved to be astoundingly consistent descriptions of a long-

Dr Kenneth Ring, psychologist at the University of Connecticut, is a leading researcher and authority on NDEs, having investigated hundreds of cases

standing phenomenon. Suddenly it was being reported more often than before because a name had now been invented for the experience and also, seemingly, because it was happening more frequently than it had done in the past.

Frankly, nobody knew exactly what to make of all these strange stories. After all, they are not, as popularly assumed, tales of life after death. By our very definition of death (as an irreversible state of lifelessness) those who return to speak of such things were at most 'near death' (hence the phrase 'near-death experience'). Possibly all that had changed was our understanding of the very last phases of life from where, in the past, people simply entered a coma from which they could never return. What happened after that – if it was anything other than total extinction of consciousness – was no nearer resolution than it had been before investigation of near-death experiences.

The first deep probe into the status of the reports came from Dr Kenneth Ring a University of Connecticut psychologist. He presented a real analytical study, finding that many of the patterns which Moody had suggested from his fairly haphazard collection of cases did still seem to apply even after Ring adopted controlled data sampling as normally used by psychologists. Indeed there was a remarkable homogeneity to the cases. They happened most often in life trauma situations – such as during surgery, in car accidents, when patients were dangerously ill, etc.

Nearly always the reports included the feeling of body and mind becoming disassociated, that is an out-of-body experience, followed by calm, peace and serenity and a desire not to return to the mortality and pain of the body. Often the reports extended to include the visual perception of a person at the attempts made by others to save them, which they viewed from an unfamiliar perspective (for example, floating near the ceiling or a few feet above the inert body). Common also were the 'life review' situations – which fitted extraordinarily well with the old wives' tale of a dying person's life supposedly flashing before their eyes. In some cases there was also a later stage when a space with a bright light at the end was seen, variously described as a tube, passage or tunnel. In one case this stage was likened to being sucked into a giant vacuum cleaner.

Again and again witnesses around the world used their own words to describe what was evidently the same experience. It was this recurring evidence, sought by science but which is absent in the field of strange phenomena, that made the NDE scientifically exciting.

Only rarely do these stages go further. Generally there is a barrier and progress seems to stop at seeing the light at the end of the tunnel. Once beyond this the witness affirms they could not have returned to their body. Some have made the journey down the tunnel and describe a land of rich colours and beauty, tantalizingly close, but ultimately beyond their reach. They see 'spiritual' beings, or even deceased relatives, on this threshold waiting to greet them. Some describe being given the choice of going on or returning. The further a person claims to go through the progression of stages, the more reluctant to come back to their body they seem to be.

Dr Ring discovered many of those who undergo a near-death experience find that their lifestyle has changed afterwards, both permanently and dramatically. In fact, even sceptics find this transformation significant. Few victims are left with any fear of death. Most say that dying is oddly wonderful. Some even report that they are looking forward to it when it happens again.

Experiencing death

George Carpenter from Gloucestershire describes the day he developed cramps in a crowded, public swimming pool.

> *I felt myself sucked under the water. For a few moments I battled against the awful sensation of drowning. Then a strange calmness fell over me. It no longer mattered whether I lived or died. My body was a thing and my mind was not a part of it. The sounds of the people who were gradually realizing what was happening just faded away. I was warm. I was happy. From a point somewhere above the water I watched it all and I really did not care. I just did not mind at all what they did to me.*

George next saw a long 'telescope' in the sky. He was being pulled through the air towards it. 'There was light at the end. Like hundreds of bright stars. I just knew that if I got there that was it. But I wanted to go.' Suddenly he saw thousands of pictures flash through his mind. They were scenes from his life screened across his brain like a crazy slide show that was presented at supersonic speed. He could take none of them in but did feel their emotion. Then, like a piece of elastic, he just 'snapped' back into his body. 'There was a terrible pain in my chest. Someone was thumping me on the side of the pool. I felt water squeeze from my lungs and all I could do was hate them. I knew that someone was saving my life but I hated them for doing so.'

This account is very much like hundreds of others that have been recorded. It was reported three years before Moody first published his results. So it was no fantasy stimulated by reading the psychologist's case histories. It was a genuine claim of an NDE before that term was invented.

Clairvoyant, Peter Lee, who reports that his clients include the late rock star Bob Marley and royalty, claimed he had a near-death experience. He says:

> *I was in Germany, and up the side of a mountain with some people. Suddenly I fell and was conscious of hundreds of little stones rolling beneath my body. I became aware of a clump of grass, or outcropping of some kind, and realized if I didn't grab it I was finished. Reaching out I caught it firmly. I was relieved. Then it gave way . . .*
>
> *As I fell all the events of my life flashed before me like a roll of film unwinding. Then I was aware of a tunnel of light stretching before me. Instinctively I* knew *this was a tunnel of* life, *not death. As I drifted along there were people I recognized who had passed away. Suddenly I was snapped back out of the tunnel, and I remember thinking: 'This doesn't feel much like heaven . . . there's pain in my head . . . my legs hurt . . .' I realized I was still alive and on the mountainside. When they picked me up I expected my body to jangle, but by some miracle no bones were broken.*

He has no idea what happened that day and is happy to entertain the possibility that it was an illusion conjured up by his brain to make death easier. But why not simply evoke unconsciousness? And how could the remarkable similarity of these cases come about?

A trip to heaven or hell?

Shirley Wood from Derbyshire, collapsed one Sunday morning and instantly knew that she was dying. However, she survived to describe a frightening form of NDE.

> *Suddenly a sharp pain hit me under my left breast. While I clutched my chest and gasped for air, my husband and eldest son guided me into the lounge and onto the settee. I struggled painfully for air for about twenty minutes, then my condition worsened and I drifted into blackness.*
>
> *Here there was no pain. Everything was calm and peaceful. I floated out of my body, as light as a feather, and drifted down into a black pit. I heard my spiritual voice say, 'Well, this is it, I'm dying. This must be death.' My voice sounded so clear and sharp in the black silence, and I wasn't afraid.*
>
> *Then I was floating upwards again, and another spiritual figure joined me, taller, grey-looking, with sunken eyes, who gently took my hand. Only the top half of its body was visible, the bottom was immersed in blackness. We began moving through a sort of valley.*

Mrs Wood's NDE was more horrific than is usually recorded. Possibly her 'valley' and 'black pit' owe their origin to religious imagery about death, heaven and hell. She continued to report how the figure became agitated and her mood changed. 'Its shoulders and trunk were bobbing up and down. I have never experienced such joy and happiness. It was indescribable. The being told me we were on our way to heaven. When we arrived the gates would open and a brilliant light would spill out enclosing us like the warm embrace of a lover's arms. I couldn't wait to get there, never giving my family a thought.'

They reached the gates but they stayed firmly shut and the entity responded in panic once again, crying, 'The light's not there!' Shirley Wood added, 'I could tell by its face it was very upset. It hesitated for a moment, as if wondering what to do next. Suddenly we were travelling backwards in a flash of speed. I shouted, 'I can't die yet, my family needs me!' There was a whistling sound, then I was back in my body, entering at the front of the head. At that very moment I came out of the blackness and heard myself gasp.'

Mrs Wood was unconscious for ten minutes. The attack was diagnosed as being caused by lack of iron in the bloodstream which resulted in anoxia, that is oxygen starvation of the brain and tissues. This is a popular reason offered by the medical profession for NDE. However, this is the first known occasion where an NDE resulted from anoxia which was medically diagnosed in a normal way. This case may prove to be very important when judging the reality behind NDE.

Shirley Wood was not a churchgoer, so this vision was not dictated by deeply held views. But it affected her. She told us, 'I am sure God allowed me to see Heaven's Gates, so I could tell others . . . It has taken over my life . . . I can't sleep at night. I want to do more things to help other, less fortunate people.'

Medical explanations

Theories were quickly proposed to account for what was being reported. One of the first was the idea that the NDE was a consequence of drugs that were being taken in by the patient's body. It is true that NDEs are often recorded by a person taking drugs for medical reasons. If all cases were related by seriously ill people or those undergoing surgery in an operating theatre this theory would be an obvious candidate. However, many cases happen suddenly when people are under the influence of no artificial agent. There seems to be no discernible difference between drug-free cases and those offered by patients under anaesthesia. A universal explanation is required for all the different categories.

The most popular theory which arose from the medical profession was that NDEs were hallucinations conjured up by the brain to cope with this desperate terminal threat to the body. A sort of automatic 'sleep mode' into which it switched at moments of extreme crisis. There are, however, some problems with this superficially attractive theory. For example, in 1983 in one case, a child of five underwent heart surgery to put a plastic valve in place. His mother and a nurse on duty at the time, recall how the boy recovered from the operation with one pressing question. 'Why wouldn't the doctors answer me when they put that thing inside? I was up by the ceiling watching.' Not only did the boy relate a procedure which he could not have seen, but he also described the artificial valve quite accurately yet he was considered too young to have been shown this before the surgery. If this child was simply recounting an hallucination, how could his description match reality so accurately.

The religious interpretation of what happens at death. Here St Benedict is escorted to heaven by angelic hosts. How do those who apparently experience the first stages of death but then return really describe the process?

Cases where the alleged hallucination offers detail that the witness cannot have known by normal methods are fairly common. In another case, a man in a Merseyside hospital in August 1984, correctly described the aftermath of a massive cardiac reaction which engulfed him following an injection of the drug Heparin. Unknown to the medical staff the man was allergic to this medication and was catapulted into a near-death experience, in both a visionary and medical sense.

The man described, to the astonishment of a sceptical nurse, the scene where a patient in the next bed alerted her and then recalled their successful resuscitation of him. One senior nurse at the hospital said she thought the account must be an hallucination, but she could not explain how it was so accurate, nor why patients in trauma situations like this always imagine themselves to be floating in the air above their beds observing everything that was taking place. She confirmed that her hospital had several such stories from patients each year, but there was an unwritten rule not to talk about them for fear of upsetting other patients.

In a case in the United States, reported by Dr Melvin Morse from Seattle, Washington, a woman rescued from death in her hospital bed by an attending physician, described seeing a tennis shoe on the ledge of the high-rise building. It was impossible to see this from any normal position and the doctor was sceptical but she did indeed find the shoe as described. As she held it out of sight she asked the patient to describe the footwear. The woman was able to do so correctly, despite allegedly only seeing this while in her NDE state.

If this were a defence mechanism of the body, taking the mind into a stress-free environment to alleviate trauma, unconsciousness would surely be the most normal response. Or, why not visions of being back at home or relaxing on a warm beach? Visualizing oneself in an unfamiliar position, hovering in mid-air, watching and hearing as doctors struggle to save your life seems far from consistent with reducing stress.

Hallucinations or reality?

In 1982 American cardiologist, Dr Michael Sabom, published the results of his work, the first by a medical doctor. He was particularly keen to test descriptions given by his own coronary patients who were reporting NDEs. He set out to check their stories against the real techniques used to revive them pointing out that most of us are only familiar with life-saving medical equipment from hospital dramas on television. These are often misleading, out of date or grossly simplified and do not reflect the complex methods used.

Dr Sabom, while convinced by the sincerity of reporters, claimed to be very open minded. His study was intended to be an objective test and he had no deep concerns one way or the other as to its outcome. If the patients described the resuscitation along the imperfect lines of medical TV series then clearly the result would be that they had not really 'seen' anything at all and were just hallucinating what they imagined was taking place. If, on the other hand, the details proved accurate, then, however awkward the problems posed, those patients must somehow have 'watched' the procedure that had been adopted to save their lives.

Sabom found that the results were exceptionally impressive. It seemed that the patients really were witnessing their return from the edge of death. They described methods, procedures and equipment he is confident they would not have seen in real life in the kind of detail that they were able to recall. These tech-

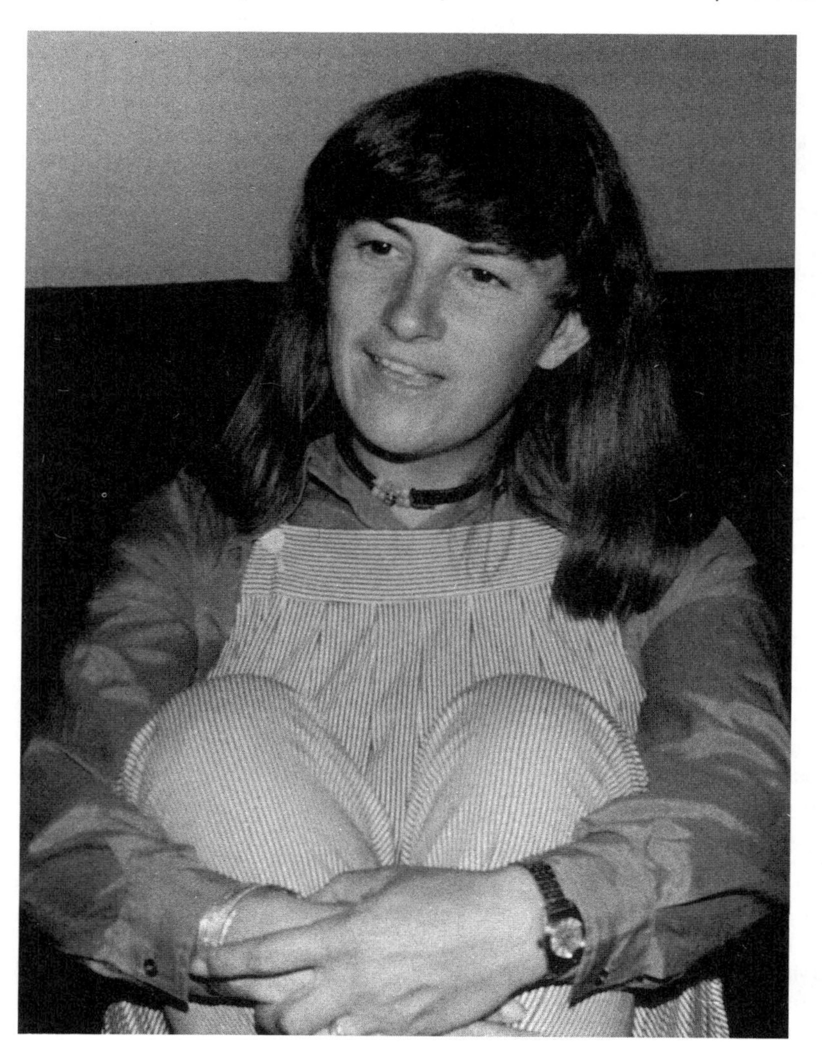

Dr Sue Blackmore, a psychologist from the University of the West of England. After her own personal out-of-body experiences she has intensively studied the NDE and produced a comprehensive 'Dying Brain theory' to explain it in rational terms which exclude the need for the survival of the mind

niques could not have been witnessed physically during the resuscitation as the patient was unconscious or even comatose. In fact Sabom said, 'The details of these perceptions were found to be accurate in all instances where corroborating evidence was available.'

The dying brain hypothesis

In 1993, Dr Susan Blackmore, an active sceptical researcher into several paranormal experiences and a psychologist who lectures at the University of the West of England in Bristol, published her intriguing solution to the NDE. She claims that all the components can be individually explained by differing processes. She terms this the 'dying brain' theory.

She argues, for example, drawing upon computer simulations of randomly firing neurons, that the tunnel and bright-light image is a consequence of how the perception system might interpret these mundane electrical signals on the retina screen at a point where bodily disfunction is causing the brain to die, for example, through serious illness. This is at a time when all other psychological context has been lost and so the image is invested with a meaning it does not have.

Blackmore argues further that the consciousness has been wrenched free of all its life associations and is merely registering a sequence of chance events taking place within its perception as it struggles to overcome the trauma imposed upon it. This lack of normal reality creates the disassociated state that seems to be experienced.

Brave as this attempt undoubtedly is, many feel that it fails to explain how any of the hallucinations could be corroborative of real events. In response Dr Blackmore states that she has tried to follow up several cases of alleged visions that were later checked against fact and in no instance has she discovered evidence that stands up. Other researchers say that they have done this more successfully.

While the authors have great respect for Dr Blackmore's work, they can only note that in their own research, they have on two occasions been able to verify with medical staff that an NDE witness did describe seeing things from an allegedly out-of-body position which they felt could not normally be seen.

David Lorimer, a former Winchester College schoolmaster and British chairman of IANDS (International Association for Near-Death Studies) told us, 'Some scientists write off the NDE as pure hallucination because their training has taught them to react in this way. We have many cases on record of NDEs occurring in people without cerebral anoxia taking place . . . We intend to begin a scientific study and publish our results in the way of articles in scientific journals.'

Lorimer describes a case where a man suffering a cardiac arrest was rushed into the emergency room of a hospital for resuscitation. Suddenly he found himself out of his body and up in a corner near the ceiling of the room. He had never been in this room before. He watched the doctors fighting desperately to revive him and heard what they said to one another. Finally they applied electric shock treatment and he 'woke up' in the intensive-care unit.

A month later this man went back to the hospital and was shown the room where he had been revived. He had never 'seen' it before in any conscious state, but recognized all the equipment, the position of the lights and so on, pointing these out to surprised staff. A number of researchers now ask how could this occur through an hallucination.

Cultural Near-Death Experiences

There are still many problems to be explained. Objective sceptical research by Dr Blackmore and others is vital.

Most of the cases on record come from the USA and Britain, which have very similar cultures. The countries have fairly identical religious backgrounds and are socially immersed within the same stimuli that comes from books, films and TV. It may not be all that surprising that the NDEs reported by the populace of similar Western societies tend to be closely related.

In order to try to rectify this imbalance, a number of attempts have been made to find out what sort of NDEs other races might produce . . . if indeed they produce any at all. So far these studies have been limited and their results conflicting, but there are at least some grounds for caution. Many of the basic parameters do appear genuinely across all cultures so far tested, but while some suspicions are verified there are also cultural variations.

One very significant difference found in recent cross-cultural studies of the NDE appears to be a religious one. Those who are brought up

Authorities

■ *Dr Susan Blackmore* is a psychologist who is the leading figure in the reductionist approach presently attempting to find a workable sceptical approach to the NDE without the need for survival of death. Although she has published reports in many sources her most comprehensive thesis is *Dying to Live* (Harper Collins, London, 1993)

■ *Dr Margot Grey* was the first British psychologist to conduct an in-depth study following her admiration for Kenneth Ring's work. She launched a parallel British research project but obtained a new database of cases and found results that were very similar. Her research appeared as *Return from Death* (Routledge and Kegan Paul, London, 1985)

■ *Dr Elisabeth Kubler-Ross* published the first book to document strange experiences of patients near death. *On Death and Dying* (Macmillan, New York, 1969). This recounts her lifetime of experience as a counsellor working with terminally ill patients.

■ *Dr Raymond Moody* with *Life after Life* (Bantam, New York, 1975) next documented a series of anecdotal stories collected during his research as a psychologist. These were the database that established the NDE phenomenon and set up future researchers. He followed this with *Reflections on Life after Life* (Bantam, 1977) which described further cases and the amazing public response to his pioneer work. Then with *The Light Beyond* (Bantam, 1988) he incorporated more recent cases and reports on the global spread of the research.

■ *Dr Melvin Morse* is a medical doctor from Seattle, Washington, USA, who specializes in researching near-death experiences reported by young children. He finds identical patterns emerging, despite the lack of social and religious indoctrination that might affect such reports from adults. His research has been published as *Close to the Light* (Souvenir, London, 1991)

■ *Dr Kenneth Ring* is a psychologist from the University of Connecticut in the USA who became one of the leaders of the new research movement after Moody's pioneering. His book *Life at Death* (Coward McCann and Geohegan, USA, 1980) was the first serious study involving detailed scientific analysis of cases. He followed this by a more philosophical work assessing the psychological impact on witnesses and society, *Heading Towards Omega* (William Morrow, New York, 1984). His most recent work has developed this concept via a major statistical survey of the lifestyles of NDE percipients compared with alien abduction witnesses seeking the striking parallels that exist. This has appeared as *The Omega Project* (William Morrow, New York, 1992)

■ *Dr Michael Sabom* was one of the first medical doctors to research NDEs among his own patients and has sought to test their visions of the resuscitation methods against the techniques that were used in those instances. His many documented cases suggest that something really was experienced and he published these as *Recollections of Death* (Harper and Row, USA, 1982)

with very strict faiths have said that they have visited 'heaven' at the other end of the tunnel. Some have even described meeting Jesus, as opposed to simply vague glowing figures or dead relatives. Others have given accounts which are more in keeping with their own religious expectations, suggesting an element of personalization does dictate the exact form of the NDE vision.

Perhaps the most important distinction are those revelations of hell, as opposed to heaven. These are sometimes reported by people who attempt suicide and then undergo an NDE. A man named Robert from North Carolina, USA, aged twenty, took fifty sleeping tablets which resulted in a heart attack followed by a coma.

Robert says, 'I was unconscious for five days . . . When I came out of it I could not shut my eyes. I found myself "travelling" . . . All I could see was the darkest black, with moving designs, geometric shapes . . . After the initial shock wore off things got better . . . Then I was moving fast through an avenue of trees. I was in a bright city.'

We may regard this as a variation on the

same theme (with the avenue of trees the equivalent of the tunnel and the bright city the same as the light seen there). However, the NDE has at least some input here clearly added by the mind of the person who experiences it. He interprets, or visualizes, the experience in a unique way from other people. Elsewhere witnesses have told of quite different variations on a similar general theme in addition to a basic pattern of real imagery common to all cases. At least part of the NDE vision is witness, or even culture, dependent.

But are we to take these accounts at face value? Have people really visited the portals of hell or the gates of heaven? Dr Peter Fenwick, a consultant neurophysiologist at St Thomas's Hospital in London and The Maudsley Hospital has studied several cases. He told Peter Hough, 'It depends whether I'm adopting a scientific or world view. Both views might be relevant. Perhaps it is all linked to quantum mechanics. The world view on survival after death might structure the reality of it – and not the other way round.'

However, this problem of just how much the individual puts into the NDE has an even deeper complication. Sometimes what appear to outsiders to be reports of an NDE are considered by the witness to be something else altogether. One of the most remarkable evaluations made by witnesses in an NDE-like state is that of a man being kidnapped by a UFO. The discovery of such an unexpected factor is only dimly perceived and will come as a major shock to some researchers.

Abducted by a Near-Death Experience

Let us consider two cases, both reported via the UFO community and believed to be 'alien abductions' by witnesses and UFO researchers (*see* page 59). What, in fact, are they?

In the first case, dating from February 1976, a Yorkshire ambulance driver named Reg lay flat on his bed, paralysed, but he was shown various images in his mind by two tall figures that appeared. They led him to understand he should go with them and suddenly he found himself floating up above the bed towards a hole that appeared in the roof. He drifted towards a 'glowing bathtub' and found that as he went 'into' this he was in a bright room where the beings began to communicate about religion – discussing 'the alpha and omega [the beginning and the end] in the Bible'. Interestingly, Reg's wife was with him in the bedroom during the later part of the experience. She saw his body looking ill and prone on the bed, but no 'aliens' were with him, nor had he 'gone' anywhere. Reg was examined by a doctor, who concluded this must have been an unusual hallucination.

In the second experience in 1964 Bob, a professional magician, went to the dentist in Cleveland. As he was due to have several teeth extracted he was given an anaesthetic to render him unconscious. At the last minute he 'rebelled' against the gas. Suddenly he floated up out of the dentist's chair and found himself free of pain, existing in a calm, peaceful void. The ceiling disappeared and he felt himself 'sucked upwards' into a big bright ball of light. Here he met two spiritual beings with robes and long white hair who conveyed a message about the future of the earth and told him he must fight to get back home. It would not be easy as demons would try to possess his body. He now saw his form down below and the dentist struggling to revive him. The dentist thumped his chest and Bob was back inside his body, feeling pain and being unwell. He could accurately describe what the dentist had said and done in his attempts at revival.

On both occasions Reg and Bob assumed that the light they entered was a UFO and the beings that they saw were aliens. Given today's cultural context it is easy to see why they would do so. Yet it is equally easy to imagine how such reports would be inferred as classic NDEs, with the tunnel, bright light and spiritual beings that were in each case triggered by illness or life-threatening traumas. Which is the correct interpretation?

Presumably, some of the detail and context we find in these cases were added by the witness because of their own beliefs and as a consequence of the space-age culture we live in. UFOs and aliens are now a dominant belief system and one might sometimes expect anomalous experiences to be falsely interpreted as a UFO event.

However, if it is true that some UFO abductions are remarkably equivalent to NDEs then what does this tell us? Surely not that people who think they have undergone a near-death trauma, such as the soldier in Vietnam, were really kidnapped by aliens?

The reverse possibility is that people who come into close proximity with energy from a

Tunnel visions

■ As medicine pushes back the point where death becomes reversible and survival from certain accidents or major traumas is possible, more and more people are claiming to have had strange visionary experiences. These occur in the period when they hovered over the abyss of personal physical extinction. Several hundred such 'near-death experiences' (NDEs) have been studied by doctors, psychiatrists and other researchers.

■ The stages of the NDE are very consistent and involve an out-of-body state, lack of pain or anxiety, being pulled along a tunnel towards a light, entering a beautiful visionary land, meeting people (often dead loved ones), making a choice to return and being back in the body racked with pain and fighting for survival. This seems to be reported from all over the world with little alteration to the sequence of events; although there are some hints of cultural bias and personalization. For instance, very religious people may 'see' Jesus or Buddha in the 'land'.

■ The further the person travels along the road towards bodily death before being revived the more these stages tend to occur. However, some of them, such as the out-of-body sensation and the tunnel of light, appear not to be at all dependent upon any medical threat to the body and are reported in cases where only fear of death is involved (e.g. mountaineers who fall long distances but land physically unharmed). This implies these are physiological, or even psychological, symptoms telling us nothing about the possible survival of the spirit.

■ The most vivid and common feature of the NDE is its transformative power. A person usually emerges without fear of death, with a much more positive outlook and with changes to their approach to life.

■ Some researchers are exploring the intriguing parallels between the NDE and alien abduction stories and are suggesting some common ground may exist between what appear to be two utterly different phenomena.

UFO could be so traumatized by the incident that a sudden NDE strikes them down through shock and the imagery then becomes confused with the preceding UFO encounter.

A third option is that some unusual but common 'altered state of consciousness' generates strange images in the mind. We as individuals evaluate these in the way that best suits our personal philosophy. For some, where the context is very clearly one of near death, the bright place is seen as a land beyond, or heaven, the tunnel effect as the journey there and the glowing forms or beings as relatives or Jesus. Expectation dictates how we interpret what we witness.

For other people, where there may not be any obvious near-death context, the experience may be evaluated quite differently. They then see the bright place as a room within a UFO, the tunnel as an energy beam sucking them inside and the entities as alien creatures.

On this premise we might, for instance, search records of fairy lore from medieval times and find some 'disguised' NDEs that were misinterpreted in that context. This would be interesting research.

The modern parapsychological view

Only recently have researchers into the NDE begun to take notice of the surprising parallels between 'their' phenomenon and the growing number of UFO abductions. The authors of this book made some attempt to interest British researchers several years ago, when they first saw the potential significance. They were unsuccessful, but fortunately Dr Kenneth Ring is now in tune with this theme and working on it.

Since 1989 he has done some excellent surveys of the interrelationship between UFO abduction experiences and the NDE, including a major comparative study known as 'The Omega Project' (an intriguing choice of name, given Reg's experience just outlined, which Jenny Randles only alerted Dr Ring to afterwards). She has discussed the results with the psychologist. He feels that they point the way towards a clear need for further research and experimentation.

Dr Ring found that UFO abductees and NDE witnesses were 'highly comparable psychologically' and that both had an enhanced sensitivity to what we might call psychic phenomena from childhood. The 'psychic track record' of witnesses was a clue some abduction researchers

Major explanations

■ *Brain anoxia* occurs when insufficient oxygen reaches the brain.This can ultimately be the cause of irreversible death in major medical crises and is known to bring about hallucinatory experiences as the brain's neurones misfire under the strain. However, some NDEs occur when brain anoxia is considered medically improbable. Cardiac surgeon Dr Michael Sabom claims to have studied NDEs in patients where his instrumental measurements at the time showed that no anoxia was occurring.

■ *Wish fulfillment* can be a powerful factor if a person believes that they are dying. They may want to escape the pain and hope to reach heaven. The heaven that they reach would be a very personalized one, in the way the NDE visions appear to be. However, why does the person always see themselves initially as floating above their body? Why not wish themselves at home with their family or in some sanctuary, such as a warm ocean or desert island? Saying that the NDE is an hallucination fails to account for both its consistency and its unmistakably vivid impact.

■ *Drug administration* is commonly blamed, especially in NDEs that occur under anaesthetic or in surgical treatment. The administered chemicals can precipitate strong hallucinations and disassociation from the body is one of the most common. However, this fails to explain why the same process occurs in cases where no drugs are administered, for example, in sudden accidents prior to, or completely without, any medical intervention.

■ *Survival of death* tries to accommodate shortcomings of other theories. The mind is claimed to detach from the body as death approaches, or thought to be approaching. In this disembodied state the spirit or soul drifts towards another reality without ties to the material world. Before absorption into the light the person may have a choice to stay or return. All NDE victims who recall being given this choice chose to come back. The main drawback of this theory is the personal nature of this 'heaven'. Some researchers argue that the realm of the mind is highly image based and so creates its own environment.

■ *The dying brain theory* as proposed recently by psychologist Dr Susan Blackmore attempts to provide a rational, materialistic compromise to resolve the various strands of the NDE without need to postulate any survival beyond bodily death. Dr Blackmore claims that as the brain disintegrates, randomly firing neurones or nerve-cells in the cortex (the outer part of the brain) can be shown by computer simulation to create a tunnel-like illusion.The bodily disassociation occurs as the mind tries to create a context in the absence of any proper incoming sensory stimuli. Natural endorphins released in the brain to deaden pain create a feeling of euphoria and freedom. As brain anoxia takes effect visual hallucinations follow that adopt a personalized pattern according to one's views of life after death. This accounts for the deviations in individual stories and the overall consistencies occur because they are descriptions of medical processes that will happen to all human beings experiencing the NDE. Those snatched back to life view the disappearance of the self or ego into a timeless, spaceless world as somehow unique. Others gradually fade into oblivion as the brain is utterly extinguished. This reductionist view does successfully explain much that is seen within the NDE phenomenon.Its main problem appears to be the conscious choice and rational thinking processes reported in the latter stages of the experience by some witnesses. All of these processes seem inconsistent with a brain on the point of complete disintegration and the almost universal sense that this is a process the individual looks forward to going through again when real death occurs. Would the complete destruction of the self and one's ego or personality really be experienced with such joy and be anticipated with such relish? It seems more likely to many that such a process would precipitate terrible nightmares, fears, phobias and distress which, in fact, almost never occurs after an NDE.

had long pointed out. Furthermore, the two experiences – UFO abduction and NDE – however different they might seem on the surface, Ring argues, 'may have a common underlying source'.

The psychologist suggests that both are symptoms of an evolution in human consciousness into what he calls the 'omega' being. Presumably, just as the body can evolve (from Neanderthal to *Homo sapiens*, for example) so too might our consciousness progress. Are those who have these experiences forerunners of what will become a more common and advanced form of mental capacity?

Dr Ring also noticed that the beliefs of those who experienced NDEs and UFO abduction were the one major difference. Those who had regarded their vision as a near-death phenomenon tended to have an increased spiritual awareness of the world. Those who had evaluated it as a UFO encounter had much more strongly formulated the view that there were alien intelligences influencing the fate of the universe. Although both sets did share a 'new age' awareness.

Possibly this connecting thread of belief is the crucial factor that makes the two experiences seem to diverge. On the other hand, the very fact that the NDE is not as clear cut or to be found only in life-threatening traumas as many would have expected, may limit its relevance to the question of our survival of death. As it can apparently pop up in something as extraordinary (and to many people, no doubt, quite absurd or imaginary) as a UFO kidnap will inevitably call into question an assessment of the NDE as some form of reality.

Dr Peter Fenwick says, 'There is still a long way to go. We need to properly assimilate information of how many people in the death crisis situation have an NDE, their psychological make-up, social background, and what medical conditions, if any, are involved.'

At present, figures suggest that less than one out of every hundred people in potential near-death situations report anything like an NDE. As such it appears to be very much the exception rather than the rule, which poses questions about its interpretation.

If indeed it were evidence of post mortem survival of the mind or spirit why do so few of us experience it? On the other hand, if evolution has found it to be a useful defence mechanism in such traumatic situations, then why is it not happening most of the time?

The case for survival

The results of a Gallup poll in the USA imply that some eight million Americans (that is one in every thirty) have encountered an NDE at some point in their lives. So far an estimated 10,000 cases have been recorded world-wide from people who claim to have some recollection of a visionary experience in the period between some sort of life-threatening situation and return to full consciousness. Studies have been carried out in over a dozen countries, ranging from America to Japan and Britain to India. There are very consistent features found in all cases; notably the lack of pain and fear, the sense of out-of-body depersonalization and the general pleasurable nature of the vision.

There seems some influence, although only relatively small, according to personal religious beliefs. The main features are still reported but are occasionally perceived according to the religious context (e.g. a strange figure might be Jesus, the Lord, or Buddha).

Studies suggest that it makes no difference whether one believes in life after death or is an aetheist. The experiences are reported just the same. But over 90 per cent of percipients emerge without any further fear of death and over 75 per cent with a belief in survival.

Tunnels and a bright light are seen in about one-third of cases, figures and rapid reviews of life images in about one-fifth of the incidents. In surveys by researchers it was noted that between 2 per cent and 5 per cent of the population report an experience resembling an NDE from some point in their lives. Most occurred during surgery (20 per cent), serious illness (45 per cent) or major trauma such as motor accidents (15 per cent).

This all makes the NDE one of the most widely experienced and yet little understood aspects of strange mental phenomena.

STRANGE BEINGS

FAIRY FOLK

Centuries before extraterrestrial beings gripped the imagination, our world was visited and populated by fairies. The breeds of such creatures were numerous, including pixies, goblins, hobs, mermaids, gnomes, hags, bogeys, elves, banshees, boggarts, leprechauns, brownies and elementals.

The word 'fairy' is derived from the Latin word *fata*, describing the individual fates of men which were personified as supernatural women. This in turn became 'fay' meaning bewitched or enchanted, then 'fay-erie' (faerie) – bewitcher and enchanter.

Fairies ranged from a few inches in height to the size of a small child or even to beings of light taller than a man. American author W. Y. Evans Wentz spent two years at the beginning of the twentieth century visiting the Celtic regions of Britain, Ireland and Brittany gathering information and first-hand accounts of fairy beliefs and encounters. One witness told him, 'There are as many kinds of fairies as populations in our world. I have seen some who were about two and a half feet high, and some who were as big as we are.'

At the head of the fairy races was the Dana O'Shee of Ireland, or The Gentry. These fairies were the aristocracy of fairyland; they were just a few inches high, headed by a king and queen and spent their time fighting, hunting and riding, with much music and dancing. Outside of The Gentry most fairy folk lived in small family groups or as solitaries. The society of fairies on the Isle of Man reflected that of Ireland. T. C. Kermode, a member of the Manx Parliament, related to Wentz a personal experience:

One October night, I and another young man were going to a kind of harvest-home. My friend happened to look across the river and said, 'Oh, look, there are the fairies.' I looked across and saw a circle of supernatural light, in which spirits became visible. The spot where the light appeared was a flat space surrounded on the side away from the river by banks formed by low hills; and in this space around the circle of light, I saw come in twos and threes a great crowd of little beings. All of them, who appeared like soldiers, were dressed in red. They moved back and forth amid the circle of light, as they formed into order like troops drilling.

A different category of fairy were the Elementals who were spirits of streams, lakes and trees. Elementals were also deemed guardians of animals and crops. There were tutelary fairies – guardians of a particular human family from generation to generation. Most notable of these were the Irish or Highland banshee. Banshee actually means 'fairy woman'. They had foreknowledge of impending deaths in the family and were heard weeping and wailing on the eve of such tragedies.

What fairies looked like

Most of us have been brought up with the image of the Disney fairy: a blonde-haired beautiful female creature with wings, wearing a gossamer dress. In fact, this is only one description of a fairy because goblins and trolls also belonged to the fairy nation. Fairy types varied, in the British Isles at least, depending on where in the country they were located. The gentle fairies, as described above, were situated in the Midlands. Mischievous but rarely dangerous, they were noted for cleanliness and their magical influence over crops. These fairies often entered the homes of humans during darkness to steal food and bathe their children.

Along the east coast of England, in the Fen Country of Lincolnshire, resided the Yarthkins. In contrast to their cousins in the Midlands, these were ugly, malicious creatures bent on doing man evil. Further north the small merry fairies were outnumbered by boggarts, bogey beasts and murderous Redcaps. In Wales there were mainly female fairies headed by a Fairy King. It should be noted, however, that fairies have the ability to shape-shift, that is alter their external appearance. Wentz was told the following during his expedition: 'They are able to appear in different forms. One once appeared to me and seemed only four feet high and stoutly built. He said, "I am bigger than I appear now. We can make the old young, the big small, the small big."'

Where fairies come from

St Augustine suggested that such supernatural beings were the fallen angels of the Old Testament working to usurp God. Many modern

Farmer's daughter, Shui Rhys of Cardigan, often associated with fairies, until one day she disappeared. Fairy tales mirror modern-day contacts with extraterrestrials. There are many stories of people being taken on trips to fairyland. Today, witnesses are abducted aboard spacecraft, and menaced by small creatures with large heads, which in another times would have been labelled goblins

Christian fundamentalists who believe in the objective existence of supernatural creatures concur with this conclusion. In Ireland however, while this explanation is accepted in part, it is understood that most fairies were angels taken in by Satan's lies, but not intrinsically evil.

When ejected from Heaven along with the evil ones, this lack of true wickedness stopped their fall to Hell, leaving them on the Earth together with man. The *Lucifugi* or followers of Lucifer fell deep into the caverns beneath the Earth, becoming trolls, gnomes, kobolds and other light-hating creatures. Others fell into the sea and became mermaids. Those that fell near men's habitations were to become brownies and hobs whose fate was to work for men without payment. Where they fell and what they became was decided by God.

Other theories of fairy origin included the belief that they were spirits of the dead who had passed away before their proper time. One school of thought held that they were intermediate beings who were placed between man and the angels. A more prosaic explanation for the origin of fairies is that they were the evolved personifications of the ancient pagan gods. Some scholars were forced to find an explanation which reflected the apparent physical dimension of fairies.

David MacRitchie, a nineteenth-century historian, argued that fairies were a pre-Celtic race of pygmies who lived in caves and were nocturnal. The new races of man would view these ancient mysterious creatures as being on intimate terms with the old gods and would thus fear and revere them. Out of necessity the pygmies would be expert thieves, taking what they wanted – mostly women and cattle – under cover of night. Some might strike up a relationship with settlers, doing odd jobs in exchange for food. Attractive though the idea undoubtedly is, it does not explain a lot of fairy lore, it lacks archeological verification, and cannot account for more modern fairy encounters.

Fairy legends

There is a rich tapestry of legend describing encounters between men and fairies. The lesson to be drawn from such tales is that even fairies who acted benignly could not be trusted. Chronicler, Walter Map, born around 1140, recorded many stories. One involves 'Wild Edric', a warrior of historical note who helped resist the Normans.

Edric fell in love with a fairy and during a visit to a house in the Forest of Clun, near the Welsh Border, snatched her away. She agreed to be his wife as long as Edric never reproached her with her origins. They were happily married for many years. When Edric at last submitted to William the Conqueror, he took his fairy wife to Court where everyone admired her beauty. In a fit of impatience, however, he asked her if she had been associating with her sisters, whereupon she vanished. Edric searched the forest for the fairy house, but it too was gone forever. He searched in vain until he died.

After his death, witnesses claimed to have seen Edric and his fairy wife beside him riding along the border. These anecdotal accounts nicely round off what seems nothing more than a romantic tale. Indeed, like other fairy legends, there are many versions based on the fundamental plot of the story. Yet despite these, the last recorded sighting of Edric and his lady was made as recently as the nineteenth century. Then, a man and his daughter, who was illiterate, accurately described the figures' Saxon costumes.

Walter Map also wrote of the legend surrounding King Herla, the ruler of Ancient Britain. The king was visited one day by a grotesque fairy riding on the back of a goat. He claimed to be Lord of all the Fairy Realms, and offered to attend King Herla's wedding, if he would return the compliment a year later. He agreed, and on the wedding day the Fairy King arrived with many equally ugly courtiers, loaded with gifts. At cockcrow they departed but not before King Herla was reminded of the agreement.

As promised, one year after his marriage, the king and his entourage set out for the Fairy King's wedding. They spent three days and three nights drinking and feasting, but at last decided to leave. The Fairy King gave Herla the gift of a small greyhound. This was to sit on the king's saddle, and the men were warned not to dismount until the dog had done so first.

They set off for home but discovered that everything had changed. There was no palace, and when the party stopped and asked a peasant the way, they could hardly understand him. In fact three hundred years had passed, and in the meantime the Saxons had conquered Britain. Some of the knights dismounted at understanding this, and immediately crumbled to dust. The king, remembering the fairy's warning, remained mounted with the rest of his men. They are said to be riding still, waiting for the dog to leap to the ground.

A common story is that of the midwife called out to attend the birth of a fairy, or at least a hybrid child. At the time the woman does not realize that the ugly old man who fetches her is a fairy. She goes with him to a house where a young girl is in labour. The bed where the birth is taking place is surrounded by small children. As the baby is born, the midwife recognizes the mother as a mortal girl who was abducted some time before from the village. She is given some ointment to put on the baby's eyes, but accidentally smears some on one of her own. At this the midwife sees her surroundings for what they really are. The cottage is a cave, the children are hairy imps and the old man is even uglier than before. She stifles her surprise and the old man pays her. Some time later she sees him stealing at the local market and accosts him. He asks her which eye she can see him with, then immediately blinds it.

A fairy hoax

People who sought proof for the objective reality of fairies thought they had found it with the case of the Cottingley Fairies. The story itself holds the key to its final resolvement.

Ten-year-old Frances Griffiths and her mother were staying with her older cousin, Elsie Wright, at her home in Cottingley, a country town in Yorkshire. Elsie was sixteen and despite the age difference, the girls got on very well together. Both girls had been absorbed by fairies long before Frances moved in during April 1917.

At the bottom of the Wrights' garden was an enchanted strip of land known as Cottingley Glen. A beck with a little waterfall ran through this overgrown wooded area. Frances was the first to see the fairies during solitary visits to the glen. Before long the girls were spending most

Elsie Wright smiling at a fairy in full flight – one of the 'Cottingley Fairy' photographs. Several photographic experts pronounced the pictures genuine. It was almost sixty years later that the women confessed. Frances claimed she really had seen fairies, and had concocted the pictures when put under pressure to produce evidence

Above The first 'Cottingley Fairies' picture, showing a line of fairies dancing before ten-year-old Frances Griffiths in Cottingley Glen *Left* The original drawing by Shepperson in *Princess Mary's Gift Book* which was used by the girls as the basis for the fake photographs

of their free time along the brook and in the trees, taking sandwiches for little picnics.

Their parents became suspicious and asked what attraction the glen held for the girls. Annie Griffiths, in particular, was furious with her daughter for the number of times she had slipped into the beck and muddied her clothing. Finally, under pressure, Frances admitted they had gone into the glen to watch the fairies. When Polly Wright asked her daughter if it was true, Elsie backed up her cousin and agreed they had both seen the little people. The adults laughed and over a period of weeks made fun of them. They decided to provide 'proof' of the sightings, so Frances persuaded her cousin to help her take some photographs. It was now July.

Arthur Wright was an enthusiastic photographer and it was with some reluctance that he allowed his daughter to borrow his camera. As far as Mr Wright was aware, Elsie merely wanted to take a picture of Frances in the glen.

They returned in a short while, and Elsie and her father disappeared into the makeshift darkroom beneath the cellar stairs. Arthur was impressed by his daughter's natural photographic skill, but not by the line of fairies dancing in front of his niece. Mr Wright presumed the figures were cardboard cut-outs as Elsie had drawn fairies for years, her talent earning her a place at art college in Bradford. However, the girls asserted the picture was genuine, and their mothers, knowing them to be truthful in all other respects, began to believe. A further photograph, this time of Elsie and a gnome, was taken in September.

The whole affair may have rested there but for Polly and Annie attending Theosophy meetings in Bradford in 1919. On one occasion a lecturer discussed the subject of fairies. Afterwards, Polly approached the man, Edward Gardner, and told him of the photographs. He asked to see the original negatives and passed them on to an expert photographer, Mr H. Snelling, for analysis. Snelling pronounced them genuine. Further he claimed they were not double exposures, neither were the figures cut-outs, nor could he find evidence of the pictures being retouched. Snelling's statement went a long way to making the photographs acceptable.

It was at this point that Sir Arthur Conan Doyle became involved. Conan Doyle, the creator of Sherlock Holmes, was a Spiritualist and a firm supporter of parapsychology, having himself witnessed some unusual phenomena including the materialization of figures out of ectoplasm. Conan Doyle was already commissioned to write a feature on fairies for the *Strand Magazine* when he heard of the Cottingley photographs. He was initially cautious regarding the validity of the pictures, but Gardner convinced him of their authenticity. This was reinforced by three more pictures taken by the girls in August 1920.

The article, published in December, was headlined AN EPOCH MAKING EVENT . . . DESCRIBED BY A. CONAN DOYLE, and included the first two photographs. The magazine was sold out within days. A second article followed in March 1921, and a book, *The Coming of the Fairies*, a year later. By now any reservations Conan Doyle may have previously voiced had vanished. Despite the army of critics, the famous author was convinced of their authenticity, as were a number of professional photographers.

As Conan Doyle and Gardner went on lecture tours promoting the pictures, the sceptics tried to duplicate the photographs to prove they were faked, searching children's picture books for the inspiration for the tiny figures. In fact, the proof was there all the time – right under the nose of Sir Arthur Conan Doyle.

Over the decades the story resurfaced occasionally. Elsie appeared on the BBC programme 'Nationwide' in 1971, and both women were interviewed four years later in *Woman* magazine. In 1976 they revisited Cottingley for Yorkshire Television. Throughout, Elsie and Frances stuck to their story. One year later, paranormal investigator Fred Gettings, discovered the proof that helped expose the hoax.

He was looking through a copy of *Princess Mary's Gift Book*, when he came across an illustration by Claude A. Shepperson. It depicted three fairy figures, and these were in exactly the same stance as the fairies in the first Cottingley photograph. The wings which had been added in the photograph were probably drawn by Elsie. The book was published in 1915, two years before the first fairy photographs. More remarkable was the fact that a contributing author to the same volume was none other than Sir Arthur Conan Doyle.

Computer image enhancement in 1980 supported the belief that the images were two dimensional cut-outs rather than three-dimensional living beings. Subsequent analysis of the photographs and the original camera by

During the search for proof of fakery, sceptic William Marriott manufactured this picture showing Conan Doyle in front of a ring of dancing fairies. He illustrated how easy it is to fake fairy pictures

expert Geoffrey Crawley showed that the first two photographs had been retouched and were not the originals. An original print was then discovered which proved Crawley's analysis. It was blurred and lacked depth. Was it Snelling, first to 'verify' the photographs' authenticity, who had carried out the work? Examination of the last three pictures convinced Crawley they had been faked by superimposing the image of a 'fairy' onto the leafy background of Cottingley Glen.

Another paranormal researcher, Joe Cooper, who had befriended the two old ladies in 1976, finally drew a confession from them that took the lid of the whole affair. It was the old story of a practical joke which grew and grew until neither of the girls could dare tell the truth. But there remained a few mysteries.

Frances Griffiths fell into the same trap as many others who have experienced supernatural phenomena. She was put under pressure to produce evidence for a phenomenon only she could perceive. She had been annoyed at being made fun of by her elders so she and Elsie set out to concoct 'proof', but failed to see the ramifications.

Frances admitted that the first four photographs were faked, but was adamant that the fifth, referred to as the 'fairy bower', was genuine. Elsie said all of them were faked, although both women claimed to have taken that last picture. Even though the 'evidence' turned out to be fraudulent, Frances stuck to her original story that she had indeed seen fairies in Cottingley Glen. It is not clear whether Elsie saw them or not, or was just a willing collaborator in the hoax. However, Elsie's obsession with drawing fairies went back many years, which perhaps hides another story.

There was one more anomaly in the second picture. That was the oddly elongated hand of Elsie Wright, as it reaches out to touch the fake gnome. Was this the 'hand' of the 'cosmic joker', having the last laugh of all, or simply a normal optical effect?

Modern fairy tales

Did people really have encounters with fairies? It is hard for us to appreciate that in their time fairies were an integral part of the fabric of pre-industrialized society. Most people accepted them as a matter of fact. Certainly poltergeist phenomena must have played its part in reinforcing this belief (and who is to say that fairies are not behind the poltergeist anyway?). Poltergeist effects are usually quite childish, but can develop ugly, dangerous traits. This is exactly the personality of the fairy – kind, mischievous, but cruel if crossed.

Peter Hough has investigated several poltergeist cases. In one, a hammer flew across a room and narrowly missed the head of a neighbour who had come round to help a family. Two men who scoffed at a poltergeist in the Bull's Head, Swinton near Manchester, were 'attacked' in the cellar during the early hours of Easter Monday 1985. One of them was taken to hospital with a head wound which needed stitches. He had 'fallen' against a beer barrel. One case which typifies the childish nature of the phenomenon occurred in a corner shop near Stockport during 1990 and 1991.

The couple who took over the premises had to carry out a lot of renovation on the old Victorian structure, including re-building the gable end. They were rewarded with odd

What we refer to as poltergeist activity was once considered to be the work of fairies. As this picture from an American case shows, poltergeists can be mischievous and alarming – the character of the little folk. The popular theory is that pent-up emotional energy can become externalized, wreaking havoc. But not all poltergeist activity is destructive, a lot of it is orderly and seems to indicate an intelligence

A former police officer took this photograph on Ilkley Moor in 1987 and then saw a silver disc disappear into the sky. It seems to show the type of creature that in ages past would have been perceived as a 'goblin' or a 'boggart'

happenings, such as discovering the contents of a packet of cigarettes which were burnt, while the packet itself was unaffected. Objects unaccountably moved in the flat upstairs but the phenomenon really got into its stride in the shop itself.

Over several mornings the couple came downstairs to discover things amiss. Several tubes of polo mints from a box were found in a neat row on the shop counter, their labels facing uppermost. Similarly half a dozen small coffee cups were found in an orderly row, with their handles all facing the same way, and several cans of coke had been taken out of the cooler and lined up on the floor. But the *pièce de résistance* occurred while the shop was locked up and empty. The owner's wife had gone out for twenty minutes on business. When she returned and stepped into the kitchen area of the shop, she could not believe her eyes: 'There, on the kitchen floor were all my saucepans. They were in a line from the smallest to the largest, their handles all pointing in the same direction like a line of soldiers on parade. At the head of them all, standing like a sergeant major, was my wok!' In times past there is no doubt that such phenomena would have been blamed on fairies, indeed *was* blamed on them.

Fairies *per se* were a dying breed when the Industrial Revolution took hold during the nineteenth century, and the blossoming of science sought to rationalize every phenomenon, natural and supernatural. Men's minds were seduced from their 'irrational' relationship with the dark forces of nature, towards the vulgar reductionist machinations of scientific thinking. Science enabled man to soar in the sky and ultimately delivered space flight. It had rejected one alien world for a host of others, across the cold reaches of outer space. With this came the return of the fairies.

On a May morning in the 1970s Mrs Kent left home to visit her daughter's house a short distance away. The canteen assistant lived on a Lancashire council estate surrounded by reclaimed land seeded with grass, bushes and trees. It was 06.15 hours, a bright and sunny day, as she walked along the road on the edge of the estate where it meets the slope of a hill. Suddenly she stopped. Just beneath the brow of the hill stood a figure staring intensely down at her. Although it resembled a man, there was something very odd about his clothes. 'He' was wearing a one-piece silver suit with a cloak tied at the neck, which had raised pointed lapels. Boots projected above the grass line, and a pointed hat sat on his head. Behind and to the right of the figure, stood a large silver spherical object which cast a beam of light down the hillside. The phenomena was still there on her return a few minutes later, but was gone when she went out at 06.45 hours for her bus to work. What was an 'elf' doing standing before a 'space ship'?

At a place called Eagle River in the United States, a sixty-year-old chicken farmer discovered a saucer-shaped object hovering outside his home. Three 'men' could be seen inside who reminded Joe Simonton of 'Italians'. They were about five foot tall with dark hair and skin, wearing blue one-piece suits styled with turtle-necked collars and knitted helmets. Joe felt no fear, and noticed that one of them was 'cooking on a flameless grill of some sort'. After expressing an interest in the food, he was given three small flat cakes. In return, they requested a jug of water, which Joe duly, and perhaps wisely, gave them. The object 'disappeared' and afterwards the local sheriff vouched for the chicken farmer's good character.

Two of the cakes were analysed in the Food & Drug laboratory of the U.S. Department of Health, Education and Welfare, at the request of the United States Air Force (USAF) who were investigating the case. The cakes contained terrestrial ingredients, but surprisingly perhaps, lacked salt. UFO researcher Jacques Vallée pointed out in his book *Passport To Magonia* that during encounters with fairy folk small cakes were often exchanged for water, and the Gentry could not abide salt.

What are we to make of UFO abduction accounts (*see* page 59)? There are now thousands of recorded cases of individuals, sometimes in groups of two or more, apparently abducted and taken onboard 'space craft' by extraterrestrials. There they undergo examination before being given a tour of the ship and taken for a ride. Finally they are returned to Earth.

This sounds remarkably like the tales of people abducted and taken away to fairyland. The physical appearance of the entities, many of them small ugly creatures, bear a resemblance to historical descriptions of trolls and boggarts. Take another look at the photograph of the 'Ilkley Entity' on page 164. This could easily be interpreted as a picture of a member

A couple who renovated an old Victorian shop near Stockport experienced a number of strange happenings. One morning they came downstairs to discover five tubes of mints laid out on the counter, with labels facing uppermost

Fairyfolk facts and observations

- Fairies are not confined to children's story books, but were believed to exist by many people in pre-industrialized society.
- Many people today believe we are being visited by extraterrestrial beings. Some psychologists state that UFO abductions cannot be wholly accounted for by abnormal psychological states.
- The parallels between fairy lore and modern UFO lore are too close not to be related.

of the lower order of fairies. In the 1950s tall, pale-skinned blond aliens predominated – very similar in manner and appearance to the angelic type of fairy.

One of the fundamental tenets of fairy lore is that of the changeling. This is a fairy baby left in place of a human child which has been abducted, or a baby which has been possessed by a fairy. The changeling has its place in UFO abduction lore too. As discussed in Alien Abductions on page 66, American researcher Bud Hopkins has investigated a number of cases where female abductees under hypnosis describe being impregnated with alien sperm. At a certain stage in their pregnancy the women are re-abducted and the foetus removed. Later they are taken again and shown their hybrid child. Many of these women can produce medical confirmation that they had pregnancies which were mysteriously terminated.

The 'aliens' which are encountered by late-twentieth-century humans seem just a modern refinement of the fairy races of the pre-industrialized world. The latest research reveals that UFO abductions are not psychological at root, neither are they folklore myths. If this refers to 'aliens' then it must also apply to 'fairies'. In modern UFO encounters, the visitors impress on the witness the dire environmental consequences of nuclear power and other poisons of mass pollution. The ecological message is a common component in these accounts. If, indeed, fairies do exist objectively in some form or another, sharing this planet with man, it makes sense that they should seek to warn him of the fatal destruction of their common environment.

THE MEN IN BLACK

Many UFO percipients claim to have been contacted by 'officials' who have sought to persuade them not to talk about their experiences. Stories of intimidation by men of sometimes oriental appearance emerged from North America in the 1940s and 1950s. They were popularly termed 'The Men In Black' (MIB) because many of them wore black business suits and drove black cars. Their clothing was immaculate and the cars, although out-of-date models, looked in mint condition.

Witnesses to some quite innocuous UFO events told ufologists how they were visited by the Men even before the sighting was reported. They sometimes claimed to be from the military or representatives of the government.

Tales of intimidation

During 3 August 1965, a man named Rex Heflin claimed to have taken a series of daytime polaroid photographs of a circular object near a Marine Corps airfield in California, Fortunately he made copies because two men who said they represented the North American Air Defence Command (NORAD), took the originals away. Later, NORAD disclaimed any connection with these 'officials'.

Two years later, scientists from Colorado University became interested and began a study of the photographs. Heflin received a visit from another group of 'officials'. They arrived late Wednesday afternoon on 11 October 1967, wearing air force uniforms. This time Heflin demanded their credentials and wrote down relevant details.

They asked about the photographs and enquired what he knew about the Bermuda Triangle – an area where shipping and aircraft have disappeared. During the interview he noticed a car parked nearby with someone

American, Rex Heflin, took several pictures of this object in 1965. They proved to be very controversial. Afterwards he claimed to have been visited by bogus officials on two occasions

sitting on the back seat. A violet glow, as if from an instrument panel, illuminated the figure. At the same time Heflin was aware of interference on his radio. Investigation failed to trace any of these men and they seemed to be imposters.

The Heflin case has its fair share of controversy, especially surrounding the photographs. However, cases such as this may have been instrumental in the United States Air Force circulating a memo dated 1 March 1967, headed: *Impersonation of Air Force Officers.* But phenomenal intimidation is not confined to the United States.

At 03.00 hours on 25 July 1952, Carlo Rossi, a 53-year-old fisherman, was on his way towards a favourite spot along the River Serchio in Italy. At a point where a high bank hides the surface, he became aware of a strange light over the river. Curiosity drove the Italian to climb the bank and peer over. He was amazed to see an enormous circular object hovering there. It appeared to be taking in water through a tube trailing from its underside. The thing hovered by means of propellers, and except for a rustle like silk, was completely silent. Carlo later gave a full description of the object.

> *I judged its diameter to be about 25 metres [82 feet]; a round disc, black with no appendages. In the centre of the disc a small turret protruded 3 metres [10 feet] below, and ½ metre [1½ feet] above. It was as big as a room, transparent like glass. Inside were visible four thin tubes attached to a large cylinder in the centre. A bluish flame shot through with orange passed from one cylinder to another. These flames lit up the disc. The craft had five propellers beneath and attached to the upper part were three more.*

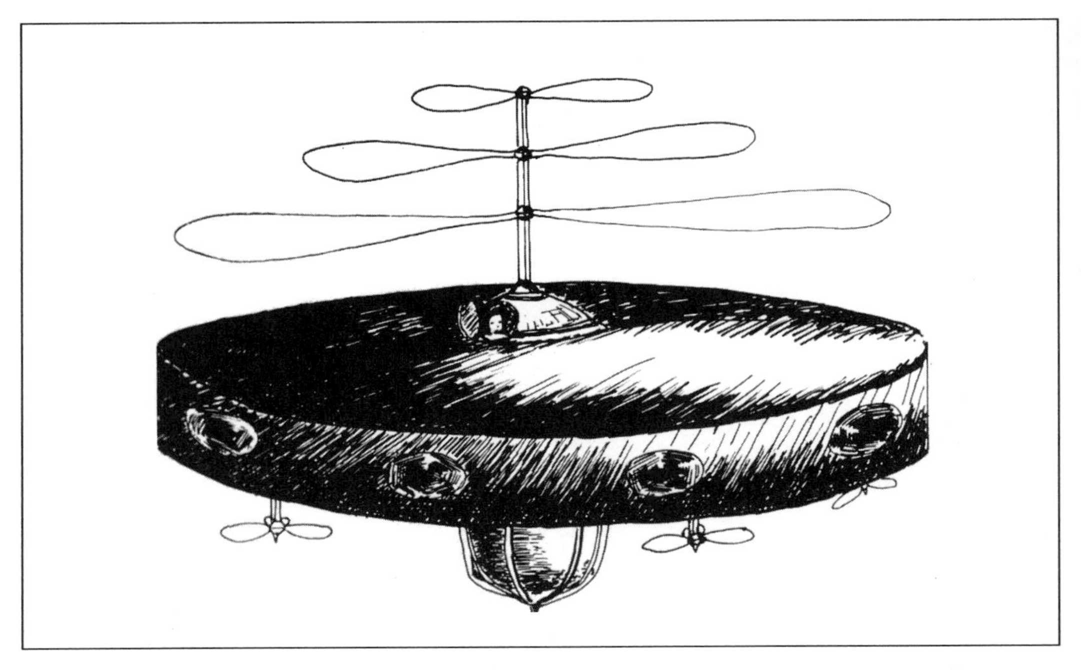

After observing this object in 1965, Carlo Rossi was later approached by a sinister foreigner in blue overalls

Suddenly a porthole opened in the upper part of the turret, terrifying the man. A human-like face peered out directly at Carlo. The figure began to indicate to someone else inside, and the Italian turned to flee in panic. As he reached the bottom of the embankment, a green ray passed close to his head sending an electric shock surging through his body. Throwing himself to the ground, he saw the object rise before disappearing at a fantastic speed towards the village of Vianeggio.

Carlo Rossi kept quiet about his experience because of the trauma and fear of being branded a liar. Two months afterwards he had an encounter with a stranger.

On 15 September at 04.00 hours, he arrived at his usual fishing place to find another fisherman already there. Carlo described the man as a 'foreigner' who was wearing blue overalls and carrying a fishing rod. He was very tall and thin with a pointed nose and grey eyes, and spoke Italian with an unknown accent. Casually he asked Carlo if he had seen an aeroplane or other object over the river. Surprised, but remaining cautious, Carlo said 'no'.

More words were exchanged, and the stranger offered him a cigarette of an unfamiliar brand. As Carlo inhaled, nausea overtook him, and he immediately extinguished it. As he was about to place it in his pocket, the cigarette was unceremoniously snatched from his fingers, screwed up and tossed into the water. Without further comment the stranger rushed off leaving Carlo bemused and afraid.

Fearing now for his life, he made a formal testimony to the authorities. The case was re-investigated in the mid-1970s by the Prato Ufological Research Group, according to researcher, Mary Boyd. Carlo Rossi was now dead, but several relatives were interviewed.

The witness's son, Sergio, still remembered the agitation of his father who kept repeating, 'They are surely not going to harm me because I saw that thing!' According to several friends of the dead man, the stranger had been seen in the village, wearing a soldier's uniform.

One of the most famous cases concerning these Men is that of American Albert K. Bender. He started up a civilian UFO organization in 1952 called the International Flying Saucer Bureau (IFSB). During the short year of its existence it spread rapidly to several countries. Suddenly Bender closed the bureau, blaming threats from an unnamed outside source. The full story emerged ten years later in a book entitled *Flying Saucers and The Three Men.* The story was incredible to say the least.

The intimidation began when a man with

glowing eyes followed Bender about town. This was followed by a visitation in the privacy of his own room: 'The room seemed to grow dark, yet I could see. I noted three shadowy figures in the room. They floated about a foot off the floor. All of them were dressed in black clothes. They looked like clergymen, but wore hats similar to the Homburg style. The eyes of all three lit up like flashlight bulbs . . .'

They warned him to cease being involved in investigating the UFO phenomenon because they, the aliens, were carrying out a secret ten-year project on Earth. In an out-of-body state, Bender was transported to a secret alien base beneath the Antarctic. By way of reinforcing the threat, the Men told him of abducted humans used in experiments, and warned that nothing must jeopardize their mission. This involved extracting a substance from sea water.

Bender, both impressed and frightened, promptly set about disbanding the IFSB, much to the chagrin of close friends and associates. In 1976, a letter appeared in the journal *Flying Saucer Review*, from a Captain E. L. Plunkett, former British representative of the IFSB. Plunkett said he had been personally informed of the closure by Bender. The affair had frightened the American so much that he had been violently sick for three days. True to his word he withdrew totally from all ufological activity.

A case which vividly depicts the bizarre behaviour of the Men (and Women) in Black, involved Dr Herbert Hopkins. He was helping out investigators in Maine, USA, during September 1976 by using hypnotic regression on a witness. One evening after his wife and children had gone out to see a film, the telephone rang. The caller claimed to be a New Jersey ufologist, and asked if he could come and discuss the case Hopkins was assisting in. He agreed, and the caller said he would come round right away. It was quicker than Dr Hopkins expected.

As he went from the telephone to turn on the outside light, a man was already climbing the steps. Hopkins said there was no way the man could have got from a telephone to his house in so short a time.

This same effect was reported in a British case in 1990 by a woman undergoing 'alien contact'. She climbed the steps from a street that was empty and entered her flat in Reading. Only moments later the doorbell rang. An MIB was there who asked strange questions. Investigators Robert France and Clive Potter conducted numerous tests and showed that it was impossible to climb the steps in the very brief time available to the visitor.

In the Maine case, the man wore a black suit, black hat, tie and shoes with a white shirt. Everything looked brand new. Hopkins thought he looked like an undertaker, but like other MIB victims his natural reactions seemed subdued, or controlled. Even when the man took off his hat to reveal he was completely hairless, and accidently smeared the lipstick he was wearing, Dr Hopkins just discussed the case as if everything was normal. After a while the visitor's speech began to slur, and he made some comment about a loss of energy, then left.

If this sounded like the doctor had suffered a mental breakdown, the theory was squashed flat by what subsequently transpired. A few days later Dr Hopkins' daughter-in-law Maureen took a call from a man who claimed to know her husband John. He asked if he and his female companion could visit them one evening. John said he did not remember the couple, but met them anyway at a local restaurant and brought them home. The couple, in their mid-thirties, wore old-fashioned clothes and were very odd, although this was not questioned at the time.

Both the visitors walked with very short steps. The woman's legs seemed oddly jointed and her breasts were very low. The visitors accepted drinks but never touched them all evening. The man asked some very personal questions of his hosts and fondled the woman, asking if he was doing it right. He asked Maureen if she had any nude pictures of herself. All in all they lacked social etiquette!

When they decided to leave, the man became paralysed. His companion asked John to help her move him, but suddenly he came to life, and the two marched out of the house in single file without even saying goodbye.

Both episodes were totally bizarre. Although there was no obvious UFO connection with the second visit, the innuendo is that all of the three MIB were robots or alien creatures of some sort struggling to act human and failing. As active field investigators, the authors have been involved in several cases where UFO witnesses have been intimidated by 'officials'.

The authors investigate

During the night of 31 August 1980, Mrs Hollins, having felt unwell and not wishing to disturb her husband, was asleep on the settee in the lounge of her bungalow in Golborne, Lancashire. At roughly 02.00 hours she was awoken by a bright light shining through the curtains. She thought the family car, parked outside, may be on fire, and so she rushed over to the window and tugged the curtains apart. The middle-aged woman was amazed to see that the light was coming from an object low in the sky, hovering over some distant trees. Intrigued, but nervous, she went outside for a better view.

She later described the object as spherical, light grey with a black band around its circumference. There were three dark blue nodules in a triangular configuration sitting in the grey area. Red flames and sparks issued from the object, and the sky was a misty pink colour. Spellbound, Mrs Hollins watched as a structural device was lowered from the craft into the trees below, then retracted after a few minutes. Suddenly, at a terrific speed, the object soundlessly moved towards her before disappearing from sight.

Fortunately, the sighting received corroboration from two other female witnesses living about a mile away. Their description differed in some respects to that of Mrs Hollins, but that might be accounted for by the different angle of viewing. It was these other witnesses who contacted a local newspaper who published a short account.

When Peter Hough first interviewed Mrs Hollins, he found a lady of average intelligence struggling to come to terms with what she had seen. She was very cooperative and friendly, providing tea and biscuits. After most of the salient points had been covered he casually asked her if anyone other than the local press had contacted her. She replied that only a week ago she had received a telephone call from a man at Jodrell Bank – the radio telescope science centre near Knutsford in rural Cheshire. Apparently he was very interested in her sighting and wished to question her in detail. When she asked how he had obtained her telephone number, he laughed and replied, 'Don't worry, it was from a very good source!' Mrs Hollins asked the man if he meant the newspaper, but instead of replying, he informed her he would be in touch shortly to arrange an interview.

Peter Hough returned on 21 September with her statement typed up and ready for signature. Apparently the man had telephoned once more. This time he had asked if she would accompany several other witnesses to Jodrell Bank where they would be interviewed. Mrs Hollins said the man had an American accent and claimed to have been in the company of (then) Senator Carter in Georgia on the night he witnessed a UFO – since identified as the planet Venus. He also warned her not to speak to 'cranks' or UFO groups, and stated he would contact her with the details.

It was becoming a little alarming. Jodrell Bank staff are not interested in UFOs – in fact they regularly pass on reports to the Manchester UFO Research Association. Mrs Hollins gave Hough what she thought was the name of the caller, and he warned her to treat any arrangements with caution. Jodrell Bank were contacted and as expected, they flatly denied any involvement, and had no knowledge of the caller claiming to represent them. There was some bemusement at the other end of the line when the caller's name was mentioned. The name Mrs Hollins had supplied belonged to the new director of Jodrell Bank. Someone, or something had 'borrowed' his name.

When she was contacted several days later she informed Hough that the man had telephoned to say he would pick her up in a car on Wednesday 8 October to take her to the science centre. He told her of the conversation with Jodrell and warned her to have nothing to do with the man. The voice had been lying.

The arrangements were for 13.30 hours. Unbeknown to anyone, Hough parked his car at the entrance to the cul-de-sac to watch and challenge anyone who arrived. Fifteen minutes past the appointed time he decided to call at the bungalow to learn if the appointment had been cancelled. There was no response to repeated knockings at the door, although an electric fire was on in the lounge, and voices from a radio or television set could be heard.

During the following week he made several attempts to contact Mrs Hollins on the telephone, but she made excuses not to speak and once pretended to be a relative. On another occasion she claimed she could not talk because there was 'a man in the house'. Mrs Hollins seemed to have undergone a personality change. She now sounded a very frightened lady and would not cooperate.

The case was rested for two years. Then in 1982, it was decided to contact Mrs Hollins again. Jenny Randles telephoned her on the pretext of doing some research for a new book.

Alfred Bender's sketch of the Man In Black who followed him around town. Bender was so frightened by his experience of MIB that he immediately disbanded his UFO organization, the International Flying Saucer Bureau

Mrs Hollins agreed to be interviewed about her UFO sighting, and arrangements were made to call round the following Thursday at 13.30 hours. As the investigators knocked on the front door, a sound like someone slamming shut a door came from the back. But it was discovered that the wood had swelled in the damp weather, and someone, in a panic, had been unable to close the door properly. Strangely, even though it was daylight, all the curtains to the back windows were closed. They went into the kitchen and found toast under the grill which was still warm. Although they were certain that Mrs Hollins was hiding in the bungalow, they decided to respect her privacy and left. When Jenny Randles called the witness later, she explained that her son had suffered a minor accident and she had accompanied him to hospital that morning.

What brought about this change in Mrs Hollins? Something had frightened her very badly. Had the voice turned nasty, as often happens, and issued threats?

The intimidators seem 'real' enough, and although generally they visit witnesses when they are alone – giving fuel to the critics who believe the phenomenon is purely of a subjective character – this is not always so. When two men bullied their way into a household in Bolton, Lancashire, to 'interview' a receptionist about her close encounter, both mother and father were present.

The pair displayed many of the classic hallmarks of the MIB. Their sinister demeanour was reminiscent of old spy films of the 1940s. One of them who called himself 'Commander', claimed to have lost an arm while serving in the RAF, but 'commander' is a naval term. The other character, with a darker complexion, hardly spoke during the entire evening, and sat with an anonymous black box on his knee, claiming it was a tape recorder, yet no tapes were changed in four and a half hours.

The Commander questioned the girl about her sighting, then tried to convince her it was a weather balloon. When this failed he told her she was a liar and she had made the whole thing up. While she was in tears, the Commander now agreed the incident had happened but warned her not to speak about it to anyone else. The two men finally left and were driven away by a third in a black car, which looked old yet was in mint condition.

Perhaps more unusual was the muted response of the father to the interrogation of his teenage daughter. The gentleman runs a business and does not suffer fools, yet he just sat there with his wife and did nothing.

Witnesses have commented at how MIB visitors seem unfamiliar with normal everyday objects. Only four investigators, including the authors, knew the identity of a former policeman who had an abduction experience on Ilkley Moor, Yorkshire, in December 1987. Further, the man produced a photograph of a 4½ foot (1.37 metres) tall creature taken just after the alleged abduction (*see* page 164).

On Friday evening, 15 January 1988, Philip Spencer was visited by two 'officials' from the Ministry of Defence. He telephoned Peter Hough just after the incident in a distressed state. They had arrived at approximately 20.30 hours and stayed for some fifty minutes. He wanted to know how they knew his identity and details of his experience. Philip wanted to keep his identity secret.

After answering a knock at the door, he was confronted by two middle-aged men wearing suits. They showed him identity cards with the letters 'MoD' printed across the top. Underneath were photographs of the men and their names. Philip clearly remembered one of them, 'David Jefferson', and the other he recalled as 'Davies'. Davies was very quiet while his colleague did most of the talking.

They wanted to interview him about his UFO experience, Jefferson explained. Philip was so stunned he never asked them how they had traced him. He related the incident but without mentioning the photograph. However, Jefferson wanted to know if any photographs existed, and Philip admitted one had been taken. When they asked to examine it, Philip said it was with 'a friend'. At the time the negative was in Hough's possession as he was about to have it examined by photographic experts. Amazingly, the men did not pursue the matter and left.

Philip thought their behaviour during the interview was very strange. For instance, they seemed unfamiliar with the electric fire and asked him questions about it. Crucially, although she took no part in the interview, Mrs Spencer later verified in two separate conversations that she was witness to the visitors.

Hough wrote to the Ministry of Defence with details of the visit. They confirmed that no one from the department had called on Mr Spencer, and further, said that the layout of the identity cards was one not used by them. However, they ignored repeated requests about whether a 'David Jefferson' or a 'Davies' worked for the MoD.

In a case investigated by Jenny Randles a couple who spotted a UFO in the early morning on the moors near Thirsk, Yorkshire, were visited by a man claiming to be from the MoD who even drove the press from the house. He asked over and over about just one detail of what the witnesses saw – a strange T shape door which glowed like a laser light source.

Over the years many good UFO investigators have quit without giving adequate reasons. Malcolm Fenwick worked for several years with British Aerospace, designing weapons systems for military aircraft. He was interested in the UFO phenomenon for its engineering possibilities because of his technical training, and joined the Manchester UFO Research Association (MUFORA) of which Peter Hough is chairman.

The night an incident occurred, Peter Hough was at Fenwick's house helping him with some research. Hough left, and the following evening received a call from Fenwick who sounded very afraid.

'Look,' he said, 'I don't want you to take this personally, but I'm quitting the subject. I don't want anything more to do with either MUFORA or ufology. I can't tell you why I'm resigning except that last night after you'd gone, something terrible happened here. I'm not going into detail but I must stress I want nothing more to do with ufology.'

It was eighteen months before Fenwick agreed to tell the whole story. 'Not long after you had gone,' he told him, 'the phone rang. I answered it and a man's voice addressed me. He spoke very clearly without accent. He knew all about me, my work with weapons systems, what I was doing then. There was no doubt he was very well briefed. What is more, he knew all about MUFORA and my involvement with ufology. That was the purpose of the call. He said I had to think about my family, and quit MUFORA and the subject altogether. Now. Then he rang off without giving his name . . .'

In several other, well-attested cases, top researchers have given up overnight, even burning their books and files. Jenny Randles was once involved in a threatening situation.

After investigating a UFO incident near a military base and writing a book about it, she was called by a man who claimed that he was a scientist with high-level clearance within the MoD. He advised her that he was checking out a theory and would call her back. True to his word he did so some days later, telling her that the matter was something he preferred not to

Men In Black fact file

■ The term 'Men In Black' came about in the 1960s after American ufologist Albert K. Bender claimed he had been harassed by sinister 'men' wearing dark clerical-type clothing.

■ In America, the phenomenon received a high profile from John Keel, the writer and investigator. He claims to have received telephone calls and harassment from beings following him in black cars. He believes the MIB are part of the UFO phenomenon. *Our Haunted Planet* (Futura 1975) is perhaps his best book on the subject.

■ The archetypal 'man in black' is, of course, Satan who was believed to ride in a black carriage pulled by a team of black horses. Modern day MIB often drive around in vintage black motor cars in immaculate condition.

■ Men In Black pretend to be 'officials', and warn victims to keep quiet about their UFO experience. They encourage people to believe they are from secret government departments, or are extraterrestrials in disguise attempting to suppress accidental sightings of their spacecraft.

■ Sceptics, like researcher Hilary Evans, agree that these incidents are not hallucinations. The MIB often interview witnesses in the presence of other people.

■ The authors have investigated many cases of intimidation by the Men In Black. Some of the encounters are bizarre, and it is hard to make any sense from them at all.

■ Although linked to the UFO phenomenon, the authors own research indicates that they can manifest in other guises too. Recent visits, principally in America and Britain, of bogus social workers, illustrate a parallel phenomenon.

delve into any further and if she knew what was good for her she would also drop the case. 'You are messing with something for which you can end up at the bottom of the Thames,' he explained.

Who or what are the men in black?

The Men In Black case stereotype, as exemplified in early American cases, is not always strictly adhered to. Researcher Hilary Evans picked out thirty-two well-documented cases for study, and found that the components of three men in black suits, with oriental facial characteristics, arriving in a black car, were not all necessarily present in any one incident. In fact, four of the sample were not visits at all, but telephone contacts. In only five were three men involved, and in the majority (twenty) just one MIB in each case was responsible.

These phenomenal intimidators usually exhibit some trait of absurdity. This can manifest itself in dress, content or phraseology of speech, or general behaviour. Witnesses have described them behaving 'like zombies' or acting 'as if drunk'. Often there is a supernatural component to the MIB. In one case the visitors did not leave footprints in freshly fallen snow.

Who or what are the Men In Black? Hallucinations, mischievous and malicious hoaxers, government agents bent on misinformation, aliens or something paranormal?

In his book *Visions, Apparitions, Alien Visitors*, Hilary Evans concludes:

> *The most probable explanation for the Men In Black phenomenon is that it is a hallucination. But such an explanation presupposes the necessary psychological mechanism whose existence has yet to be demonstrated. And it does not account for the uniquely life-like nature of the MIB entity. MIB percipients do not simply have a convincing sight of their entities; they engage in a complete and consistent series of actions – they open the door for them, talk to them, offer them drinks, often over an extended period of time.*

As active investigators in the field the authors can conceive of no psychological mechanism which could fulfill most or all of these criteria. In addition to the above, as we have seen, the MIB are often perceived by more than one witness. In our view it is ludicrous to attempt to attach a psychological explanation to these cases.

The hoax hypothesis could be an explanation for some incidents. But who are the hoaxers? Most MIB contacts, whether by telephone or personal visit, possess 'inside' information about the witness. How could a hoaxer obtain this information?

Are they government agents out to intimidate witnesses to suppress sightings of the UFO phenomenon? This supposition was given some credibility when the American Freedom of Information Act came into operation in 1978. The United States government had for many years informed investigators that its interest in UFOs had finished with the closure of Project Blue Book – an official investigation which sought to debunk the subject – in 1969. Documents released under the new act proved that both the FBI and the CIA had maintained their investigation of UFO incidents after this date.

Some ufologists, mainly American, believe that the bizarre behaviour of the MIB is proof that they are extraterrestrials trying to suppress information of spacecraft movements, through the use of intimidation. They point out that the unusual behaviour and unfamiliarity with everyday human artifacts is indicative of alien beings trying to act human but falling short because they lack the social nuances and correct behaviour patterns. The 1940s gangster appearance has been chosen because of the sinister portrayals of this stereotype in old films. Indeed, to extraterrestrials, the gangster image would be ideal for their purposes. Unfortunately, the aliens fail to understand human culture enough to realize the absurdity of adopting such outmoded motifs.

What of the paranormal option – that the MIB are a branch line of an ancient archetype, customized for the UFO phenomenon? Antiquity is littered with legends of tall dark men visiting people to issue threats. Montague Summers, the writer, for instance, discovered in the early twentieth century that on 2 June 1603, a court in France heard the following account.

A young country boy, claiming he was a werewolf, admitted abducting and eating a child. Apparently, 'The Lord of the Forest' forced him to carry out the crime. He described the entity as a tall dark man, dressed in black, who rode a black horse. Often Satan was said to manifest himself as tall dark man who rode in a black carriage pulled by four black horses. This juxtaposes very well with the Men In Black, who arrive and depart in black cars.

PHANTOM VISITORS

Two major scandals in Britain have caused parents to lose confidence in the Social Services system. The first was the Cleveland Affair. In Cleveland, 121 children were diagnosed as having been sexually abused in the first half of 1987. These children were taken away from their parents on 'evidence' produced by Dr Marietta Higgs and Dr Geoffrey Wyatt. An enquiry ruled that the tests produced ambiguous evidence and most of the children were returned to their parents.

Not long after this debacle a new child abuse phenomenon started to gain momentum in Britain after its inauguration during the early 1980s in America. The term 'Satanic ritual abuse' (SRA) found itself into the vocabulary of social workers after a series of seminars organized by Christian fundamentalists primed by their American counterparts.

Cases of 'ordinary' child abuse developed a devil-worship dimension when youngsters were found to exhibit some of the signs of SRA. This list was drawn up by a prominent Church social worker, and included bed wetting and the singing of obscene nursery rhymes. Children were encouraged to describe their abuse by witches and satanists, and to name other children who were also victims. The result was the mass removal of children from many towns and cities including Liverpool, Manchester, Salford, Nottingham, Derby and South Ronaldsay in the Orkneys. The cases took months to get to court and when they did, the majority of children were returned home because social workers and the police could not produce any evidence.

The full truth about Cleveland and the satanic ritual abuse scandal may never be known. But the effect on many parents, particularly those who lived on large estates, was a fear and loss of confidence in the social services system. They had discovered that without any evidence, social workers could take their children from them for several months until the cases went to trial. Have these fears produced a hybrid of the Men In Black – the Phantom Social Workers?

Bogus officials

This new phenomenon swept Britain in the middle of the SRA debacle, beginning in December 1990. Police received hundreds of reports of callers presenting themselves in some kind of 'official' capacity and trying to procure very young children. They claimed to be social workers, health experts or from child protection groups like the National Society for the Prevention of Cruelty to Children (NSPCC). Yet their behaviour was so ridiculous that no one was ever likely to be taken in by them. Although opportunities did arise for the abduction of children, the visitors never did so. There is only one known case where a mother allowed a couple to take away her two young sons. The boys were returned 45 minutes later and told police that the couple had simply taken them to a park and bought them ice cream.

Phantom social workers often arrived in pairs. Many of them looked 'official' and carried clipboards. In one incident a man and a woman called at a house and claimed they were from a child-minding service, and asked the mother if they could pick her children up from school. When the mother asked for identification the pair just walked away. Quite often parents described a bewildered look on the faces of these 'officials' when challenged, as if it deviated from a set script, and they were then unable to cope. Others called claiming they were investigating child abuse, and in a minority of cases children were examined. There was a report too of a woman seen taking photographs from a road, of children playing in a park.

In a bizarre incident, a woman claiming to be a social worker called at a mother's house and said she would have to take away her two-year-old daughter because the woman had spotted the child eating a jam sandwich. She said the girl should not be eating jam at that time and a tug of war ensued between the 'social worker' and the mother, as she attempted to take the child away. The mother won.

Incidents still occur. In September 1992, parents from four different addresses on an estate reported a visit by a couple asking about their children. During February 1993, an Oldham mother claimed she was visited by a woman dressed in black, who claimed to be from the Social Services Department. Needless to say, she too turned out to be bogus.

Crimebuster Police fear child abuse

BEWARE OF BOGUS

Photofits of the phoney officials. Women pictured centre and right are probably the same person

DON'T LET THESE WOMEN INTO YOUR HOME: Photofits of the phoney officials. Women pictured centre and right are probably the same person

POLICE are warning parents all over Britain to be on their guard after a spate of attempted child snatches by a gang posing as child care officials.

Already there have been more than 14 attempts to abduct children in this way. At least one involved a serious sexual assault.

So far the attacks have been confined to the Yorkshire and Humberside area, but police say the gang could strike anywhere.

The evil trio – two women and a man – pose as health care officers or representatives of the National Society for the Prevention of Cruelty to Children and call at the houses of mothers with young children.

They ask to examine the children, often stripping them naked and touching their private parts. Then they tell the shocked parents that the children have been sexually abused and need to be taken into care.

So far, the ruse has not succeeded. No parent has allowed their child to be taken away. But it may only be a matter of time before the gang is successful.

The catalogue of terror began on January 30 at Park Hill flats, Sheffield, the home of 24-year-old mother Elizabeth Copeland.

Two smartly-dressed women, both carrying briefcases and notebooks, arrived at her council flat and asked if they could examine her two daughters – Samantha, two and Dancia, five months – to ensure they were being well cared for.

Since then, the same gang has been linked to a number of similar incidents.

On February 8 at Athesley North estate in Barnsley, a three-year-old boy was examined by a woman claiming to be a health visitor. The mother asked for ID and the woman produced an identity card with a picture of someone else. When challenged, she ran off.

Another incident took place on March 9 at Central Drive, Rossington, Doncaster, when a woman asked to examine an 18-month old girl. When the mother refused, the bogus official said she would return with a warrant. She didn't.

On March 22, a woman claiming to be from social services approached the father of a six-month-old boy in Thorne, near Doncaster, and asked to examine him.

When the boy's father sensibly asked if he could call social service bosses to confirm she was genuine, the woman refused and ran off.

Detectives believe the gang may have taken to wearing disguises to avoid being recognised.

On two occasions, the first at Birdwell, near Barnsley, the second at the Wyburn estate in Sheffield, a woman stopped passers-by to ask the names and whereabouts of young children in the area, obviously trying to uncover likely targets for the next strike.

"We don't know what the motives of this gang may be," said a police spokesman. "It is possible that they plan to steal a child to keep for themselves, but it is also possible that they may get some sexual gratification from intimately examining these young children."

It is thought the man may hav some kind of Svengali-like hol over the women, who are trying t abduct the children on his behalf.

Police have not ruled out th possibility of even more sinist motives, such as procuring the ch dren for pornography or devil wo ship rituals.

The most serious atta occurred on March 16. Two wom claiming to be from Rotherh social services entered a house a ing to examine a one-year-old bo

They claimed they had infor tion showing the child had b abused. They stripped the baby in front of his mother.

The "examination" which lowed was so intimate that de tives have classed it as a ser sexual assault.

Chief Superintendent D Foss, in charge of the investiga

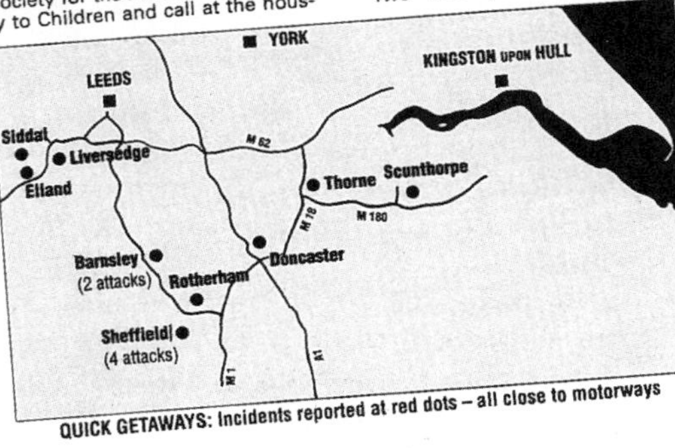

QUICK GETAWAYS: Incidents reported at red dots – all close to motorways

Are phantom social workers an off-shoot of the Men In Black phenomenon? There are now hundreds of cases on file.

gang is behind mystery calls on mums

SOCIAL WORKERS

CHECK THEIR I.D: Reconstruction shows how well-dressed, well-spoken women have approached mums with small children

She wanted to examine my baby son

JULIE HARGREAVES, a mother in her late twenties, had a lucky escape when a smartly-dressed woman called at her home near Barnsley, West Yorkshire, earlier in the year.

"I let her into the house and we stood in the hallway. She asked me how old Jordan was. I said he would soon be one. She then asked if she could do his yearly check, which was quite feasible, as it was due," Julie told Radio 4's Today programme.

"I wasn't suspicious at first," she said. "She didn't have any cards, but she was very well-spoken, very smartly dressed, very confident in the way she spoke.

"She had all the mannerisms of a health visitor. She was very friendly.

Lucky

"I said I'd rather she didn't do the examination there and then as Jordan was upstairs in bed. She kept on insisting and I kept on refusing. In the end, the woman lost her temper and stormed out."

Julie was lucky. Had she allowed the woman to examine Jordan, he, like several other innocent children in the area, would have been intimately examined by a stranger who had no qualifications and no right to do so.

At first, Julie blamed herself for being duped, but she now realises that she was not the only one. Any parent, off-guard for a few minutes, could easily be taken in.

"At first I felt very stupid that I'd allowed myself to be caught out, but when you study the situation, I'm sure anybody would have done the same," she said.

"I'm just glad that Jordan was in bed and not downstairs, otherwise I probably would have allowed her to examine him."

● If you have any information which you believe could help catch this gang, call Rotherham police on the number below and ask for the incident room.

varns all parents to be on their uard. "Always ask for identity ards or call the office the official is upposed to represent. If the visitor genuine, he or she won't mind."

He also warns parents not to be oled by smartly-dressed, well-oken people who know both ur name and the names of your children – this information can be easily gained from neighbours.

"We are treating this matter very seriously. It would take only one moment of carelessness for someone to be fooled. The consequences of that don't even bear thinking about," said Mr Foss.

The male gang member is described as being between 35 and 40 years old, around 5ft 8in tall with a distinctive black square ring on his right hand.

One of the women is between 35 and 40 years old. The other has dyed blonde hair and is in her late twenties.

TONY THOMPSON

If you have any information, call police on (0709) 828182

33

Despite extensive police investigations, no one has been arrested in connection with these disturbing visits

Official theories

The police and the media initially concluded that the calls were made by members of a paedophile ring. Why such people would run the risk of approaching parents in broad daylight and continue to do so while a national police investigation was under way, was never explained. It was also theorized that the bogus officials were really thieves, casing houses for future burglaries. Detectives then hit on the idea that the mysterious callers were 'sex vigilantes' out to trap parents they suspected of molesting their children, and belonging to a Christian fundamentalist group. Detective Superintendent David Foss admitted the police were baffled: 'It isn't clear that they are paedophiles and it is not clear that the motive is theft. You would have thought that would have happened by now. But equally, they have had plenty of opportunity to abduct a child, and this has not happened either.'

Despite the fact that officers from a dozen forces investigated the reports, not one person to date has been arrested in connection with the visits.

A case investigated

Peter Hough interviewed one mother who claimed she had been visited by phantom social workers. The incident occurred on Sunday, 6 December 1992, in Leigh, Lancashire. Mr Carter, a self-employed car mechanic was out working, leaving his wife, a community nurse, and her two young children at home alone. At approximately 10.30 hours there was a knock at the door. She answered it to a man in his early forties holding a clipboard. He was thinning on top, wore a moustache and had a scar on his right cheek. The man introduced himself as 'Albert Sutcliffe', and Mrs Carter said he was from either Adlington or Accrington Social Services Department – she could not remember which. He sounded Scottish, and produced an identity card which displayed a photograph. Mr Sutcliffe claimed he was following up reports that Mrs Carter was not feeding her daughters properly. At this point he called to a female colleague who was standing near the silver Mercedes car the couple had arrived in. As she came up the path, a white van drew up with three women sitting in the front. They all wore scarves, as did the woman who now joined Mr Sutcliffe. Mrs Carter noticed the words 'Child Protection', emblazoned on the scarf.

Mr Sutcliffe told Mrs Carter that the van was used to take away children, but in her case he had decided it would not be necessary, and instructed his colleague to tell the driver to go. He requested that they be allowed inside the house to check that everything was in order. Bewildered and upset, she invited them in.

The woman who was in her early thirties was smartly dressed like Mr Sutcliffe. She spoke very quickly so it was often difficult to understand what she said. They checked the food cupboards, then were shown upstairs where the children were playing. Mr Sutcliffe asked if they could examine the six-year-old, but her mother declined, and he accepted the refusal. Downstairs Mr Sutcliffe told Mrs Carter she could keep his card because it had a telephone number written on it where she could contact him. They asked to see the girls' birth certificates and the woman took some details from them.

Mrs Carter stood up to go into the kitchen to make some coffee. As she did so, Sutcliffe asked for the card back as he wanted to write something on the back. The card was never returned. After they had finished their drinks, the couple left. The visit had taken about 45 minutes in all.

Mrs Carter was convinced by Mr Sutcliffe and his colleague, and it was only when she talked it over with her husband that he realized there was something very wrong. The police became involved and they interviewed Mrs Carter, took away the birth certificates for finger prints and made door to door enquiries.

Neither Peter Hough nor the Carters' know the result of the police investigation. Peter Hough wrote twice to the police with the couple's agreement, requesting answers to several questions which might throw light on the mystery. But to date the letters have not been acknowledged.

A check with some of the neighbours drew a blank. No one remembered seeing either the Mercedes or the white van, although one man thought he had seen the car since, cruising slowly past the house. Were the Carters picked out at random? When Peter Hough asked to talk to the eldest daughter, his request was politely turned down. Some time previously there had been an unsavoury incident involving the child, and they did not want her upset by the current event.

The truth is that we only have Mrs Carter's word that anything happened at all. However, her husband verified that his wife was very upset on his return, although she had already washed up the tell-tale coffee cups.

The parallels between these bogus officials and the Men In Black are obvious. There is always a touch of the bizarre. In this case the scar and the name 'Sutcliffe', notoriously linked with the convicted killer of thirteen women, seemed designed to intimidate. Child molesters, burglars and religious extremists would not drive around in an easily recognizable car. This must lead us to conclude that the Men In Black and Phantom Social Workers are one and the same phenomenon, and therefore, the MIB have nothing to do with extraterrestrials.

Strangers on the sidewalk

The weird phenomenon of bogus social workers visiting and examining children before vanishing into the nether world from which they have come seems a peculiarly British problem. But that is not entirely the case. There have been several examples of strange visitors appearing on the sidewalk outside American homes.

At Council Bluff in Indiana in April 1992 a male and female visitor, much as in the typical British pattern, arrived at a house and examined two children for bruises. They did nothing further but said they could take the children away if they wished to do so. The workers, of course, were later checked out and found to have no official existence.

The same was true of two women who arrived in southern New Jersey the following month. They visited several homes claiming to be 'case workers' and having false ID cards to prove this. A three-year-old girl in one house was stripped and examined for bruises. They warned of future visits but never returned. Subsequent enquiries revealed them to have no real status. Nor, as usual, could they be traced by the police.

It is worth recalling that both the Men in Black and Satanic Ritual cases began in the USA (in the 1950s and early to mid-1980s respectively). From here they spread to Europe. The bogus social worker pattern is reversed, starting in Europe and later reaching the USA in increasing numbers. The comparisons between all three phenomena are sufficiently clear to argue that there is common ground – probably a single root phenomenon variously interpreted in relevant cultural terms.

Possibly the Cleveland scandal set an expectation in Britain whereby these visitors took on the new guise. These events do seem to follow trends. Indeed, in the USA from the late 1980s visits by strange-looking clowns were reported. They, too, had no real substance when checked out. As yet they have not spread elsewhere, but experience suggests they may do.

Several American psychologists have begun to look for patterns between phenomena such as these and claims of visits to a bedroom by alien entities who then 'examine' children. There does seem a definite continuity; although what the consequence of that fact is remains undefined. Like the aliens, these bogus visitors appear utterly real to the percipients at the time but never leave any proof of their existence behind.

Official fact file

- Calls by phantom, or bogus, social workers were first reported in the 1980s in Britain and then America. They seem to be a direct response to a lack of confidence in social workers due mainly to the satanic ritual abuse fiasco and the Cleveland scandal.

- Phantom social workers call during the day and deliberately draw attention to themselves with their bizarre behaviour. Although they threaten to take children away, none of them do so.

- There are obvious parallels between Phantom Social Workers and the Men In Black. Despite national police investigations of many of these incidents, no one bas been caught. It is a complete mystery, which makes it clear that the Phantom Social Workers and Men In Black have the same origins.

THE ALIEN ZOO

MAN-MONSTERS

In the supernatural world one of the most consistent phenomena is the sighting of man-monsters. These are primarily located in the Americas, Canada, China, Africa and Russia but sightings have been reported elsewhere including Australia and Scotland. The descriptions vary, but witnesses are certain that they have seen something of flesh and blood which is neither man nor beast – an unknown hybrid of both.

Bigfoot

Bigfoot is the name usually given to man-beasts in the north-west United States, notably northern California, Idaho, Oregon and Washington's Cascade Mountains. This area includes thousands of miles of unpopulated forest. It has been suggested that only one in ten sightings of Bigfoot is reported. The authors' experience investigating the many areas of the paranormal endorses that opinion. Fear of ridicule and bemusement forces most witnesses into silence. In Bigfoot terms, this means there could have been up to 10,000 sightings since the end of the nineteenth century.

One of the earliest recorded incidents took place on the slopes of the Orestimba Peak, northern California, in 1869. A hunter on returning to his camp discovered that items had been disturbed. When this happened on several more occasions, he lay in wait, and witnessed an extraordinary sight. Here is an abridged version of his account:

It was in the image of a man but it could not have been human. The creature stood fully 5 feet (1.5 metres) high, and disproportionately broad and square at the fore shoulders, with arms of great length. The legs were very short and the body long. The head was small compared with the rest of the creature, and appeared to be set upon his shoulders without a neck. The whole was covered with dark brown and cinnamon coloured hair, quite long in some parts.

As I looked he threw his head back and whistled, and then grabbed a stick from the fire. This he swung round until the fire on the end had gone out, when he repeated the manoeuvre. Fifteen minutes I sat and watched as he whistled and scattered my fire about. Having amused himself, apparently, he started to go, and, having gone a short distance returned, and was joined by another – female, unmistakeably – when both turned and walked past me, and disappeared in the bush.

What we see here is a common feature of Bigfoot: a head that sits on shoulders giving the appearance of no neck. Did Roger Patterson and Bob Gimlin film a Bigfoot near Bluff Creek, California in 1967? They were out hunting the creature after a number of sightings in the area. Examination of their film taken on 20 October, which depicts a figure covered in dark hair moving into a wooded area, has naturally aroused controversy. Critics suggest that the film shows a man dressed in a fancy-dress costume. Certainly if that is the case, the man was approximately 6 ft 8 in (2 metres) tall and around 440 lb (199.5 kg) in weight.

Sasquatch

Sasquatch mainly haunts British Columbia in Canada, but sightings bearing his name have also been made in North America. Sasquatch is really the more northern appellation for Bigfoot for there is no real difference between the two creatures. This man-beast has a historical basis in that the native Indians recorded stories and incidents involving a 'wild man of the woods'.

One native Indian recounted to his relatives how three Sasquatch saved his life at the beginning of the twentieth century. He was acting as a guide in the Mount Shasta region in Tulelake, California, when, alone, he was attacked by a rattlesnake. He tried to find help but fainted. When he came round he was surrounded by three Sasquatch of 8–10 feet (2.4–3 metres) tall. They had made a cut in his wound and extracted some of the venom. Two of them carried him down the mountainside and left him where he could be found.

Stories abound of giant ape-creatures on the small islands off the coast of British Columbia. Witnesses claim to have seen the creatures swimming between the islands where primitive conditions still prevail. Probably the most famous Sasquatch story is that of the capture of a hunter called Albert Ostman.

In 1924 Ostman was in an area of British

A frame from the 1967 cine film of Bigfoot taken by Roger Patterson at Bluff Creek, Northern California. Critics have suggested that the figure is a man in a monkey suit

Albert Ostman, who claimed he was abducted by a Bigfoot in 1924. It carried him off in his sleeping bag and he found himself effectively held captive by a Bigfoot family. He escaped when one of them swallowed the contents of his snuff box. A tall story or a remarkable experience?

Columbia where a gold prospector had been mysteriously killed. It was where the Indians spoke of large mountain creatures covered in hair, who left tracks 2 foot (0.6 metres) long. He camped alone and set about hunting and searching for minerals. He arrived back at camp to discover that some creature had been searching through his belongings. At first he assumed it was a porcupine, but there were more incidents, and he became less sure. Ostman went to sleep with his rifle beside him, and awoke trapped in his sleeping bag, with the sensation of being carried over rough ground.

After about three hours he was dumped on the ground. He crawled out of the sleeping bag, and was able to make out in the darkness a group of four large hairy creatures. As daylight flooded the landscape he was able to see they were two adults a young male and female. The 'boy' equivalent in human terms was between eleven and eighteen years of age, 7 foot (2.13 metres) tall and weighing about 300 lb (136 kg). In his sworn statement Ostman described him thus: 'His chest would be 50–55 inches (1270–1397 mm), his waist about 36–38 inches (914–965 mm). He had wide jaws, narrow forehead that slanted upward round at the back about 4 or 5 inches (102 or 127 mm) higher than the forehead. The hair on their heads was about 6 inches (152 mm) long. The hair on the rest of their bodies was short and thick in places.'

The older male was considerably bigger and more powerful looking, with a hump on his back. The females were proportionally the same. Ostman was kept in their company for almost six days, not exactly kept captive, but not

Reports of Yetis have always interested the world's media, as in this depiction of an 1954 sighting by several sherpas with the Japanese Everest expedition

Tangible evidence? Are these Yeti footprints recorded in the snow? There have been many sightings of the man-beast by responsible mountain climbers, and footprints have been discovered on several occasions

allowed to leave either. He ate some provisions the creature had brought from the camp. Ostman's chance for escape came when the adult male snatched a box of snuff from him, and swallowed the contents. As the giant creature rolled in agony, and the rest of the family milled about in confusion, Ostman slipped away, eventually to make his way back to Vancouver.

Yeti

The legend of the Yeti covers the entire Himalayan mountain range, from China to northern Burma. Tales of enormous creatures half man, half beast – were living folklore hundreds of years before Westerners learned of the phenomenon at the end of the nineteenth century. It was in Victorian Britain that Major L. A. Waddell published his book *Among the Himalayas.* Major Waddell described the discovery of large footprints in the snow of north-east Sikkim, and referred to native stories of 'hairy wild men'.

Further discoveries were made in 1921 when Colonel Howard-Bury and his party attempted to climb the northern face of Mount Everest for the first time. They saw several figures walking across a valley, and were assured by their Tibetan porters that they were members of an ancient race of creatures, which translated into English as 'abominable snowmen'. At 23,000 feet (7010 metres) the party came across a path of enormous footprints, which Colonel Howard-Bury rationalized as having been made by a wolf.

Since then there has been a preponderance of sightings and the discovery of footprints too large for any animal, many of which have been photographed and preserved in plaster.

In 1974 a teenage Nepalese girl called Lakpa Sherpani, was grazing a herd of yak on a mountain pasture near Everest, when she claimed it was attacked by a Yeti. The creature covered in dark hair, was between 4 and 5 feet (1.2 and 1.5 metres) tall, with thick fingers and long nails, its heels turned inwards. The Yeti killed five of her yaks by twisting their heads. Lakpa was attacked too, but fortunately, just knocked unconscious. What the purpose of the attack could have been was not apparent. Afterwards police discovered footprints and handprints in the snow.

In 1992 a new expedition set off in search of the Yeti, headed by a 73-year-old Russian called Dr Zh. I. Kofman and a French documentary film-maker, Sylvain Pallix. The team, including ten scientists, was intending to search for evidence of the creature over a three-month period, but what they found – if anything – has not been reported. However, the publicity did generate fresh reports, and photographs of large footprints taken by Polish geophysicist, Andrzej Zawada. Sceptic Dr Bogdan Jankowski, who examined the prints was convinced they had been made by 'a weird animal of some sort'.

This is a sketch by biologist, Ivan T. Sanderson, of the Yeti, based on eye-witness reports

A Lamaist monk displays a reputed Yeti scalp and skeletal hand. Evidence such as this always proves ambiguous yet there must be *something* behind the hundreds of sightings of man-beasts. Do man-beasts and other 'alien' animals slip into our world from another plane of reality?

The Russian 'Bigfoot'

Chief among Russian cryptozoologists is researcher and investigator Maya Bykova. She graduated from Moscow Agricultural Academy in 1955, and since 1972 has organized a dozen expeditions to find animals unrecognized by science, and is the author of three books on man-beasts. In issue 1 of the Russian paranormal magazine *Aura-Z* published in March 1993, she recounts three recent sightings.

The first involved a professor of medicine who in the autumn of 1989 went into a marshy forested area to collect some herbs and mushrooms. Dr N. Aleutsky and two companions travelled by boat and moored up on the bank of a river. The professor left the others onboard, then walked a short distance and began collecting mushrooms. Suddenly a bear cub came up to him and yelped. At the same time he could hear the mother roaring nearby. Dr Aleutsky afterwards commented: 'I had a knife but that was poor defense against an angry beast. I let go of the basket of mushrooms and ran to the boat. Suddenly I heard a blood-chilling scream from behind. Turning my head I saw a gorilla-like creature holding the she-bear in its hands. The beast was 8 feet (2.5 metres) tall, its body covered with thick brown fur. It was female, her large teeth were bared. Holding the bear by the hind legs she tore the animal in two without any visible effort.'

The entire episode lasted only seconds. The professor's companions were not eyewitnesses and found it hard to believe the shocked academic. However, all three fell seriously ill after the episode and decided initially to say nothing of the affair.

The second case occurred in the village of

Sosnino situated just a short distance from Kargopol. It began when two creatures invaded the barracks of an army road-construction unit. One beast was about 8 feet (2.5 metres) tall and the other half that size. Witnesses thought they were mother and child. The small man-beast jumped onto a table, while its mother stood by a stove waving her arms, before fleeing back into the surrounding forest.

Both before and after this major incident other soldiers had sighted the creatures in the vicinity of the barracks. Those witnesses in the barracks became mysteriously ill afterwards. One man, probably through shock, lost his power of speech for several days. Footprints were discovered outside in the snow together with some tufts of hair which remain unidentified despite examination.

The most recent sighting took place in early November 1992 in the Dmitrov District of Moscow. Anatoly Dobrenko was walking his Alsatian dog near the children's hospital where he works when the animal began to snarl. About a hundred metres away Anatoly saw a creature covered in rusty coloured hair walking towards the forest. The witness's son, an army captain, went to the location with some employees from the hospital and found some well-preserved footprints recorded in mud. They measured 19.7 inches by 5.7 inches (50 cm by 15 cm).

The story was passed on to a local newspaper and journalists found other witnesses who had seen this creature. They also discovered evidence in a derelict cottage that two beasts had slept there and two sets of footprints were found nearby, one set smaller than the other.

Other hominoid giants

In the late seventeenth century a Spanish expedition hunting for gold in the Andes of South America, reported that they had fought and killed fourteen giant man-beasts. As more Europeans invaded the area, many stories emerged of encounters with giant 'man-apes'. The *Mono Grande* is the most established of these beasts, sighted extensively in Ecuador, Colombia and Venezuela.

Rex Gilroy, an Australian researcher, has collected over 3000 sightings of a giant hairy creature sighted across the continent. The Aboriginals have known of the beast for years, and call him by many names, including 'Yahoo' and 'Duligahl', but he is popularly known as the Yowie.

There are barely any contemporary accounts of giant man-beasts in Britain which is not surprising considering its large human population and relatively small land mass. However there are legends of rock-throwing giants, and a number of medieval church carvings and manuscripts which depict man-beasts carrying clubs. Wales has its legends, as does Scotland where the best known is about the Big Grey Man of Ben MacDhui. Over the years, a number of climbers on the mountain have experienced a sudden inexplicable fear. The most famous story was related by Professor Norman Collie concerning his experience in 1891. He was followed by something hidden in the mist as he descended from the cairn. Other climbers have described a huge hairy man-like creature with long arms and legs.

The search for the historical Bigfoot

Despite the multiplicity of legends describing giant hairy man-like creatures, researcher Michael T. Shoemaker claims that sightings of Bigfoot are largely a twentieth-century phenomenon. In issue 5 of the American magazine

Hair raising facts

- Beasts which seem to be a hybrid of the ape and modern man have been encountered in most areas of the world. So established are the sightings in some places, often over centuries, that the creatures have earned specific names.
- Sightings have been made by explorers of high repute. Tracks have been photographed and casts made. In rare instances still and moving film has been taken.
- Despite peripheral evidence, no one has produced the unambiguous remains of any of the creatures. This causes sceptics to dismiss the entire phenomenon as being without foundation, and merely an example of modern folklore.
- Some researchers cite certain cases where encounters have been of a bizarre nature as proof of the supernatural origins of the beasts. Are the Yeti, the Yowie and Bigfoot creatures from a parallel world which occasionally impinges on ours?

The former USSR is rich in sightings of man-beasts. Here, snowman hunter Igor Bourtsev displays a cast he made of a footprint discovered in 1979

Strange Shoemaker recorded the results of his examination of many historical sightings. He concluded that several were made-up stories or newspaper hoaxes, and where a genuine 'creature' was involved, it was probably a crazed hermit rather than an unknown species of man-ape. As far as Indian traditions were concerned, he concluded they were based on a mythological being grafted with animal attributes. Shoemaker believes that man-monsters are a cultural phenomenon and have no basis in reality.

Supernatural man-monsters

Cryptozoologists try to keep the subject of man-monsters on a firm physical footing. They believe that such creatures represent the fragmented remains of an evolutionary cul-de-sac. Man-beasts could well be the missing link between apes and modern man. They cite plaster casts of footprints, thousands of eyewitness reports and filmed evidence as proof of their existence.

Yet the sceptics ask, if Bigfoot and his cousins really do exist, then where are the remains? Even if they bury their dead, or cannibalize the remains, by now at least some bones should have been found. Where pieces of fur and tissue have been presented as evidence, analysis results have always been ambiguous. There is an obvious supernatural element in some man-beast encounters and the cryptozoologists cannot reconcile this.

In August 1972 a member of a family called Rogers, living in a trailer at Roachdale, Indiana, noticed a bright object over a nearby cornfield. Later, a large broad-shouldered creature was seen going into the field after noises had been heard near the trailer. The creature was covered in hair, and left a rotten smelling odour – a feature noticed in many other cases. Despite more sightings, the beast never left any tracks, even in mud, and it became so insubstantial that sometimes witnesses could see through it. Local farmers complained of up to two hundred chickens which had been slaughtered and their remains scattered about. Here we are reminded of other strange animal mutilations (*see* page 100). In one instance the beast was caught red-handed in the doorway of a chicken house. Despite being fired upon at close range, the thing moved off apparently unaffected.

Other cases demonstrate an immunity to gunfire, and some witnesses talk of being confronted by one of the beasts, only for it to disappear before their very eyes. Up to that moment, witnesses are convinced they have been dealing with a flesh and blood creature.

Are man-monsters, in common with other 'alien' animals, supernatural, and if so, by what mechanism? As we see from the above account, some beasts are seen in conjunction with UFO sightings. Are these creatures genetic or robotic constructs released by UFO occupants to draw attention away from their clandestine activities? What of man-beasts which vanish or become transparent in front of witnesses? Are they creatures from a parallel Earth – another dimension, who accidentally slip into our world, then slip back again? It is hard to accept that they are cultural hallucinations. These creatures are really here – they leave footprints, cause damage and kill chickens. Could this be the explanation for all alien animals? Speculative though it is, it provides an explanation for the lack of unambiguous physical evidence.

SEA SERPENTS AND MONSTERS OF THE OCEAN

The vast oceans of our planet hold as much mystery as the unfathomable depths of outer space. Indeed it has been argued that we know more about the galaxy than our seas. On the face of it this sounds absurd, but the oceans present as strange an alien environment as the cold vacuum of outer space.

Water covers over 70 per cent of the Earth's surface. The oceans were formed as soon as the atmosphere was stable and the planet cool enough to contain them. Their average depth is 2¼ miles (3.6 km), but measurements of up to 7 miles (11.2 km) have been recorded. If we put this into perspective, there are taller mountains beneath the sea than the highest on land. Mount Everest rises to almost 5½ miles (8.8 km). There is three hundred times more 'living space' in the ocean than on land. It is in the oceans that life first began about 600 million years ago during the Cambrian period with the emergence of worms, jellyfish and trilobites. It is currently believed that their ascendents crawled out of the seas developing into air-breathing land creatures.

When the first mammals roamed the Earth during the Triassic Period, some 225 million years ago, their aquatic cousins hunted in the seas. These included the ichthyosaurs and later the long-necked plesiosaurs. Scientists estimate that dinosaurs became extinct at the close of the Cretaceous Period about 65 million years ago. But can we be certain that in the vast unexplored alien world of the deep oceans, small communities of marine dinosaurs are not still breeding? After all, the existence of the giant squid was only confirmed as recently as 1857. In 1896 the giant carcass of an octopus, including tentacles 200 feet (61 metres) across and weighing 7 tons (7112 kg), was washed ashore in Florida. More recently in 1938 a Madagascan fisherman trapped a 5-foot (1.5 metres) coelacanth in his nets. This fish was said to have been extinct millions of years ago. In 1984 a fishing vessel called *Helga* netted a member of a new species of shark. 'Megamouth', as it became known, is 15 feet (4.5 metres) long.

A monster tradition

Sailors and Viking explorers long held a tradition of sea monsters – indeed they mounted carved serpent heads onto the prows of their ships to ward off the real thing. Sightings of huge long-necked creatures were commonplace and ancient sailors endowed them with a monster mythology. The Remora was said to suck at the keel of a ship and take from it the sexual proficiency of its crew. A fabled sea beast called the Kraken was reportedly the size of an island. There are passages in the Old Testament which refer to Yahweh's conquest of a sea monster, the ancestor of the Great Beast of Revelation.

In the past many explorers and scientists have supported the idea of gigantic unknown creatures of the sea. Sir Joseph Banks, the botanist and naturalist, who sailed around the world (1768–1779) with Captain Cook and was president of the Royal Society, had a firm belief in these creatures, so too did naturalist Thomas Huxley who died in 1895.

Historians cite mirages and sailors' obsession with death as the inspiration for monster sightings. This might be an acceptable deduction if such sightings were of historical note only. The world was a more wondrous place before the soulless age of reductionist thinking. People expected to encounter supernatural beings, demons and monsters of other realms. Why then, in an age where science has sought to impose its rationalization on the supernatural world, do sightings continue to be made of animals which patently bear more of a resemblance to the dinosaurs of long ago than to any modern species? Witnesses of high repute continue to report 'mythical' animals in close detail, thus flying in the face of scientific rationalism.

On Sunday, 3 September 1882 Mr F.T. Mott in the company of Mr W. Barfoot, a JP, and a solicitor, Mr F.J. Marlow, were standing on the pier at Llandudno, North Wales, when they saw a black snake-like creature swim across the mouth of the bay towards the Great Orme. The observation lasted two minutes and during that time they estimated the creature's length as around 200 feet (61 metres). The account was published in the prestigious journal *Nature.* Mr Barfoot answered criticisms by stating his wide experience of the sea, and conjecturing that the phenomenon the three men had observed could not be accounted for by known species of sea life.

While historical sightings of sea serpents serve to demonstrate the continuity of the phenomenon, it is more modern accounts, still continuing today, that provide us with a detailed and immediate picture of the monster experience.

All at sea

Just before the outbreak of World War II the captain of a tanker called *British Power* had a close encounter while travelling north towards Abadan in the Persian Gulf. It was a hot day and Captain Kingston-Lewis was taking a siesta on the poop deck. He stood up for a while, gazing over the side of the ship when a creature emerged from the water. Its head was horse-like, perched on the end of a neck which was 6–8 feet (2–3 metres) long. Kingston-Lewis could make out a large body just below the waves, and after about half a minute it ducked its head beneath the water and disappeared, after giving the ship a long inquisitive look.

When war broke out, a troop ship on its way to evacuate the British concession in Shanghai was the location for another sighting. As the ship raced through the South China Sea, Alfred Peterson, a male nurse, was taking his usual pre-breakfast jog around the deck when something made him stop. Initially he thought he could see a large floating tree, until he realized it was keeping pace with the boat. Peterson described the 'body' as being about 25 feet (7.6 metres) long finishing in a tail. Its colour was deeper than grey but not quite black, more the colour of a hippopotamus. In 1986, Peterson described what happened next to BBC journalist, Fergus Keeling. 'It turned and its neck stood up which was like a giraffe's, with a giraffe-like head. I could see two ears or horns similar to drum sticks. It disappeared, came up again, played then went on. It wasn't vicious, it wasn't splashing, it looked a big gentle thing.'

The war years were remarkable for a large number of encounters with unknown marine animals. A man called Welch recounted a sighting while on lookout duty on a small troop ship sailing from Durban to Bombay. The large black object he saw in the distance reminded him of an enemy submarine so he sounded the alarm bells. While the crew was panicking an officer trained a telescope on the object and announced that it was not a submarine, probably just flotsam. As the ship drew closer, however, it was obvious to those onboard that they were observing a living creature. 'It was definitely something swimming. It was like a serpent, 20–30 feet [6–9 metres] long, as thick as a tree trunk with its back arched in several places. I couldn't make out its head which was surrounded by waves.' Welch then described how the beast was eventually lost from view.

A depiction of a giant squid encountered by a French naval vessel in 1861 off the Canary Islands

Coasting

Many sightings of sea monsters are made by observers situated near coastlines. One area in particular which seems to attract monster activity is the west coast of Canada – particularly in the Strait of Georgia between British Columbia and Vancouver Island. Over centuries it has provided cryptozoologists with many well-authenticated cases. Before the arrival of white settlers, the native Indians already had a tradition of monster sightings. The beasts have been spotted so often since that they have earned the collective noun of 'Cadborosaurus' or 'Caddy'.

In 1984 a fisherman situated in Vancouver's outer harbour was confronted with a creature less than 200 feet (61 metres) away from his boat. Mr Thompson said it happened on a calm wintry day when there were no other boats nearby. He happened to glance over to the south-west and there it was. Thompson said it had obviously seen him first. It was moving between 12 and 15 knots out towards the Pacific Ocean, its head craned almost backwards, looking straight at him.

Yet again we have a description of the neck and head similar to a giraffe's. The beast was tan coloured, covered in fur and had a white stripe down its briskett and a black snout. There were two floppy ears, one which hung over its face while the other stood upright, and possibly a pair of stubby horns. Despite its giraffe-like appearance, it moved its 18-foot (5½ metres) long body through the water by undulating it in an up and down motion, like a serpent. The animal submerged into a swell, still staring back at Mr Thompson.

Caddy was sighted on New Year's Day 1985 by a Mr Cole who was standing by his lounge window looking out to sea. He was munching from a plate of sandwiches when he saw, against a wall of mist, what he took to be a man standing up in the prow of a rowing boat. He thought what a dangerous and stupid thing to do, especially in the current weather conditions, and picked up a pair of binoculars. He could not believe his eyes. 'Low and behold it wasn't a boat it was a creature! It had a dinosaur head – small in proportion to its body, a columnal neck which curved up from a swan's body which petered down to a tail.'

This creature was sighted by the crew of HMS *Daedalus* in the Atlantic Ocean on 6 August 1848

He estimated the neck to be between 5 and 6 feet (1.5–2 metres) long, the body 12 to 14 feet (3.5–4 metres) long, some 3 feet (1 metre) of it projecting out of the water. Mr Cole turned away and shouted to other people in the house, but when he looked back the creature had gone.

Monster states

The coasts around the United States of America have had their fair share of sea monster encounters. One of the most impressive occurred one afternoon in October 1983 near Stinson Beach, north of San Francisco. Members of a highway construction crew were the main witnesses. They were linked by two-way radios. Marlene Martin was safety inspector for the Californian Department of Transport, and explained what happened. 'The flagman at the north end of the job site hollered, "What's in the water, it's coming real fast, it's coming right at us!"'

Construction worker Matt Ratto focused his binoculars in the direction indicated and noted a mysterious animal 100 yards (91 metres) off shore. Truck driver, Steve Bjora, described it as similar to a huge eel, travelling at a speed around 50 miles (80 km) an hour. Marlene Martin told Fergus Keeling what happened next. 'By now it was by the shore in the rocks, in a big wave and lay motionless for five seconds or so. Then it stretched out, and I concluded it must be about 100 feet [30 metres] long. It submerged, made another wake then took off so fast I couldn't believe it. I grabbed the binoculars as it threw its head right out of the water. It thrust itself up and down, making big movements, coils and humps of its body – splashing everywhere.'

Through the binoculars Marlene was able to see its teeth, which were all the same length. The beast was described by her as a 'snake-like dinosaur'. The head was box shaped, 'similar to how people draw dragons, but the nose was shorter'. A feature which Marlene still finds difficulty in talking about, is the creature's eyes. 'The eye, although I can't swear to it, looked like it was red – a deep ruby red colour. It stunned me, we were so shocked. Never in my wildest dreams could I have imagined anything so huge, so fast. I thought: this is a myth!'

The beast was being followed by a flock of birds and there were also several sea lions close by. Witnesses were not sure whether the creature was chasing them, or they were following the monster. There were other witnesses too. The day was overcast and there was some rain. Despite this, teenager Roland Curry was on the beach and he also saw the animal. When interviewed, he admitted that he had also sighted it briefly the week before. Young Roland was one of several surfers operating off the Santa Ana River when later that same week 'a long black eel' appeared just 10 feet (3 metres) from his surf board.

On the opposite coast of the United States, lying exactly between the same lines of latitude, lies the location of that country's most famous sea monster: 'Chessie'. Chesapeake Bay cuts through the District of Columbia and part of Virginia. It covers an area of almost 200 miles (322 km) and has produced many monster sightings over the years.

One of the most detailed reports came from Clive and Carol Taylor who live at the mouth of the Chester River which runs into Chesapeake Bay. The father and daughter were walking along the beach one evening when they were aware of a mysterious wake about 50 yards (46 metres) off shore. It was 20.20 hours, and the water was mirror calm with no boats about. They speculated that it might be a dog swimming, but as they followed it along the shore line, more and more of the animal was exposed to them. First one hump and then another appeared, until a total of five were evident. The beast changed course towards the beach then swam out parallel with the shore again. On the second occasion it did this, Carol ran towards it to get a better view. She got within 30 feet (9 metres) of the beast. 'I saw a creature which appeared to be a snake between 25 and 30 feet [7.5–9 metres] long, as big around as a telephone pole, dark brown or green, a head shaped like a rugby ball, but larger.'

Her father recalled that the animal saw him first, but he kept still and it did not seem to view him as a threat. 'I saw the highlight of his head as he turned to look at my daughter. The eye looked like a large snake eye. It had a light yellowish tinge to it and was about a couple of inches long. I still didn't believe it – I stood there thunderstruck!'

Clive Taylor felt it wanted to come ashore, and speculated that this is what it did at night in order to feed.

In May 1986, a dentist called Dr Bishop who also lives on Chesapeake Bay, was visiting a friend who had just purchased a new boat. They were chatting on the dock when a snake-like head poked up from the water. 'As it did so I saw the sections behind the head come up at the same time. There were three of these. It swam by undulating up and out, up and out and was a reddish-brown colour.' Dr Bishop was later told by several of his patients of their own sightings of Chessie.

British sea monsters

There are two coastal areas in particular around Britain that have attracted the attentions of the sea-monster phenomenon. One is near Falmouth off the Cornish coast, and the other is Barmouth on the west coast of Wales.

As elsewhere in the world, the phenomenon has been sighted so often in the sea near Falmouth that it has been given a name. Locals call it 'Morgawr', which is Cornish for 'sea giant' – an appellation first used by local journalist Noel Wain in 1975. Sightings go back to at least the nineteenth century and probably beyond. Harold T. Wilkins in his book *Strange Mysteries of Time and Space*, records his sighting made at 23.30 hours on 5 July 1949.

> *Myself and another man saw two remarkable saurians, 19–20 feet (5.5–6 metres) long, with bottle-green heads, one behind the other, their middle parts under the water of the tidal creek of East Looe, Cornwall, apparently chasing a shoal of fish up the creek. What was amazing were their dorsal parts: ridged, serrated, and like the old Chinese pictures of dragons. Gulls swooped down towards the one in the rear, which had a large piece of orange peel on his dorsal parts. These monsters – and two of us saw them – resembled the plesiosaurus of Mesozoic times.*

Sightings, or at least reported sightings, increased from the mid-1970s. In September 1975 a local woman and her male friend spotted an anomaly at Pendennis Point. Mrs Scott and Mr Riley later described a large lumpy creature with bristles along its back. Its head boasted a pair of stumpy horns. The creature disappeared beneath the waves for a few moments, then resurfaced with a conger eel in its mouth.

There have been many sightings off the coast of Cornwall by local fishermen. One sighting of note was by John Cock of Redruth and George Vinnecombe of Falmouth. They were fishing 25 miles (40 km) south of Lizard Point one morning at the beginning of July 1976, when Vinnicombe spotted what at first he mistook to be an upturned boat. He alerted his friend who came up on deck, and both men realized the 'boat' was in fact a living creature. Vinnicombe said, 'I've seen all kinds of whales – pilot wales, porpoises everything that swims. Here was something different to what we'd ever seen before!'

They closed down the engine and drifted to within 20–30 feet (6–9 metres) of the beast. The sea was flat and visibility was extremely good. They were able to obtain a very good description, as George Vinnicombe told reporters.

> *It looked like an enormous tyre about 4 foot (1 metre) up in the water with a back like corrugated iron. It had lumps on the top like prehistoric monsters have. We must have woken it because a great head like an enormous seal came out of the water. It was a beautiful day – we could see a lot more of it under the water, so it must have been several tons in weight. It just turned its long neck and looked at us then very slowly submerged.*

Apparently the body was black and around 18 feet (5.5 metres) in length. The head was more of a grey shade.

Curiously, another fisherman similarly called George Vinnicombe was involved in creating monster headlines a generation before. In May 1926 he and Mr W. Rees were reported in the *Cornish Echo* as capturing a 'sea monster'. The harpooned beast was probably a 15 foot (4.5 metres) bottlenosed whale or a basking shark. In 1932 Mr Vinnicombe hit the headlines again when he harpooned 'a 35-foot [10.6 metres] Greenland basking shark off Kennack Point'.

The mystery of 'Mary F'

On 5 March 1976 the *Falmouth Packet* published two photographs of a sea monster supposedly taken during the first half of February. They were supplied to the newspaper with a letter bereft of an address but signed 'Mary F.' Researchers Janet and Colin Board examined first-generation copy prints and found them convincing. In her letter, Mary complained that a haze on the water and sunlight shining directly into the camera prevented the pictures from coming out more clearly. In her statement, she said that the part of the beast which was visible above the water was between 15 and 18 feet (4.5–5.5 metres) long. 'It looked like an elephant waving its trunk, but the trunk was a long neck with a small head on the end, like a snake's head. It had humps on the back which moved in a funny way. The colour was black or

'Morgawr': a Cornish sea monster supposedly photographed by 'Mary F'. But was Morgawr really a lump of plasticine, and Mary F. a wizard named Anthony Shiels?

very dark brown, and the skin seemed to be like a sealion's.'

Mary said the creature frightened her, that she 'did not like the way it moved when swimming'. The sighting lasted for just a few seconds.

Anthony 'Doc' Shiels, a professional 'wizard' and controversial figure in the paranormal community, has had his share of monster sightings. He has seen and photographed both Nessie and Morgawr. Shiels explains his 'luck' in the belief of a psychic link between the phenomenon and the observer, which 'makes' it manifest. Critics prefer a more prosaic explanation, certainly Shiels enjoys courting publicity. But to be fair, on two occasions at least, he has been in the company of other witnesses.

His first sighting was with his wife Christine and four of their children. They were on Grebe Beach the Sunday morning of 4 July 1976 when Shiels kept getting a glimpse of something in the estuary. His children were next to see it, and then it was spotted by his wife who described 'a large, dark, long-necked, hump-backed beast moving slowly through the water'.

On 17 November 1976 Doc Shiels had his second sighting, while in the company of David Clarke, editor of *Cornish Life* magazine. Clarke was taking pictures of Shiels on Parson's Beach, Mawnan, for a feature. Shiels drew his companion's attention to 'a small dark head' poking out of the water. They observed the phenomenon in more detail when it moved to within 70–80 feet (21–24 metres) off the rocky beach. Clarke wrote later: '. . . the greenish black head was supported on a long arched neck, more slender than that of a seal . . . The head was rounded with a "blunt" nose and on top of the head were two small rounded "buds" . . . at one point a gently rounded shiny black body broke the surface . . .'

Shiels added that the beast seemed no more than 15 feet (4.5 metres) in length. They agreed that the snail-like head possessed short horns or stalks. Both men had cameras and took photographs of the creature. Shiels's camera was unsophisticated and had a wide-angled lens which was inadequate for the job. Clarke had fitted a telephoto lens on to his 35mm camera and should have produced some excellent results. He made several exposures, but, as he allegedly discovered later, the mechanism was not winding on properly, and he finished up with a triple exposure of the 'animal'. The shots prior to this, taken the same morning, were all right. In paranormal investigation, camera equipment is noted for failing at the critical moment.

A monstrous hoax?

The American magazine *Strange* published a series of articles during 1991 and 1992 which linked Tony 'Doc' Shiels with the 'Mary F' pictures and several lake monster photographs credited to himself and other persons. Editor Mark Chorvinsky and other contributors claimed that Shiels had faked the pictures by mounting a plasticine serpent onto a piece of glass then photographing it in front of a watery background. This included two colour photographs supposedly taken at Loch Ness in May 1977. A third, almost identical picture was allegedly later handed to Shiels by a mysterious woman, in the tradition of 'Mary F.'

Critics asked how Shiels could have copyright on the Mary F. photographs when 'she' had supposedly sold rights to 'an American gentlemen'. *Strange* printed transcripts of tapes sent by Shiels to an American colleague called Michael McCormick, where the magician outlined plans for pulling off a monster hoax. During the investigation other, previously unpublished, 'Mary F.' pictures came to light. McCormick claimed he had assisted Shiels in faking these. Shiels admitted to investigators that he was responsible for these unknown photographs, but that he had taken them in an attempt to discover how the original Mary F. pictures might be hoaxed.

Shiels, in convoluted replies, would not admit to faking the other photographs, although the evidence seems by some to be very damning. Certainly no serious researcher would now consider them as strong evidence of water monsters.

The Barmouth dragon

Even in the 1990s, the former fishing port of Barmouth is still a beautiful Welsh, seaside town which has changed little over the years. Barmouth Bay boasts miles of beautiful sandy beach – the scene for many monster encounters over the years.

Legend describes an incident in October 1805 north of Barmouth in the Menai Straits. There a ship was chased, and caught, by a serpent-like creature. It crawled up the tiller and coiled itself around the mast. The crew attacked and drove it overboard. Undeterred, the beast continued to follow the vessel until a wind slowed it down and it was lost from sight.

What the truth of this tale might be, reported in an old Welsh book called *The Greal*, is hard to say. Certainly aggressiveness seems absent from the majority of other sea monster reports.

In 1971 two holidaymakers from Colwyn Bay were walking along the beach at Llanaber, when they came across some very odd footprints. These were at the water's edge and between 12 and 18 inches (304–457 mm) in diameter. Four years later, during December, Mr Holmes of Dolgellau reported finding prints in the sand near the Penmaenpool toll bridge. These were larger than a dinner plate and Holmes had the impression that they were webbed.

There were several good sightings during 1975. Six local school girls saw a creature on Barmouth beach. It was just growing dusk on 2 March when the twelve-year-olds saw it rise up and disappear into the sea. The creature was about 200 yards (183 metres) away. Carys Jones and Julie Anderson said it was about 10 feet (3 metres) long with black patchy, baggy skin. It had a long tail and neck, and huge green eyes which were visible above the waterline as it made off. The girls said its feet were like huge saucers with three long toe nails. Julie Anderson's father, a coastguard, said the girls were very upset by what they had reported.

In the summer of that year a couple sailing their sloop a few miles north near Harlech, sighted what they thought at first was a seal playing in two car tyres, half a mile (0.80 km) away. As they drew closer Marjorie and Vernon Bennett toyed with the idea that perhaps it was a huge turtle after all. But the closer they came, the more the couple realized they could not recognize the animal. Mrs Bennett described it thus: 'It had a free moving neck, fairly short, rather like a turtle's, and an egg-shaped head about the size of a seal's. Its back had two spines, which were sharply ridged, and it was about 8 feet [2.5 metres] across and 11 feet [3.4 metres] long, although the ripples on the water when it dived indicated it was probably about twice that length.'

Sea monster theories

We must ask the question whether witnesses are really encountering creatures foreign to modern science, or merely observing known animals under unusual conditions? Do they unconsciously exaggerate size and read archetypal monster motifs into a phenomenon which probably has a more prosaic explanation?

Sheila Anderson, a specialist in seals and sea lions, and zoologist Martin Angel, concerned with the ecology of the deep ocean, were invited by BBC Radio 4 in 1986 to explain some of the sea monster sightings reviewed here. Angel was convinced it was all down to misperception and estimates of size which were grossly exaggerated. For instance, in the Morgawr sighting above, the two experienced fisherman said they got within about 20 feet (6 metres) of the beast and estimated it was about 18 feet (5 metres) long. Angel supposed they had really observed a five-foot-diameter (1.5 metres) leather-backed turtle. Maybe large turtles could be an explanation for some sea monster sightings.

The leathery turtle's traditional habitat is the Caribbean and Malaysia – thousands of miles away from chill North Atlantic waters. However, turtles have been seen in the sea off Britain, and as far north as Scotland where in 1975 some fishermen were surprised to discover they had caught a specimen. One was in fact washed ashore near Mousehole, Cornwall in 1985. It was almost 7 feet (2 metres) long and weighed nearly 7 cwt (356 kg). The animal had choked to death on a plastic bag. Speculations about changes in the Gulf Stream have also been voiced as an explanation.

Martin Angel resorted to a familiar 'explanation' for eyewitness descriptions of humps and extreme lengths of water beasts. He thought otters or dolphins swimming in a line could be the answer. Again, his sceptical but more reasonable colleague disagreed. Sheila Anderson remarked that the probability of getting a number of animals in a neat row on several occasions, in different places, so perfectly that people are convinced by it, was remote.

Not all zoologists are antagonistic. One of the most detailed works on the subject *In the Wake of the Sea-Serpents*, was written by a Belgian zoologist named Dr Bernard Heuvelmans. It was Heuvelmans who in the late 1950s coined the term 'cryptozoology' – the science of hidden or unknown animals. He analysed 358 hardcore sightings which remained unidentifiable. These

In March 1975, six twelve-year-old girls saw this beast on the beach at Barmouth, a Welsh seaside town. A Coastguard remarked that the girls were genuinely upset by the experience

Sea monster facts

■ The first scientific book on sea monsters was published in 1892. *The Great Sea Serpent* was written by Dutch zoologist Anton Cornelius Oudemans.

■ In June 1973 a contemporary report of sea monster sightings compiled by Paul LeBlond and John Sibert of the Institute of Oceanography at the University of British Columbia, Canada, was published. *Observations of Large Unidentified Marine Animals in British Columbia and Adjacent Waters* divided descriptions into three categories: big-eyed long-necked types with a body hidden below the surface, a small-eyed long-necked variety with a body showing above surface, and a snake-like serpent.

■ The International Society of Cryptozoology (ISC) was founded in January 1982, in Tucson, Arizona, USA. Its president is Bernard Heuvelmans. The aims of the society are to investigate, analyse, publish and discuss all matters related to animals of unexpected form or size. or unexpected occurrence in time or space.

■ Despite, perhaps, the unpalatable possibility that sightings of some sea serpents represent a paraphysical source, the society has in its membership many respectable oceanographers and marine biologists.

■ To date no one has come forward with the unambiguous remains of a sea monster. However, the society has had some success in identifying other so-called 'hidden' animals. A long-legged puma-type cat called the onza has at last been verified in Mexico, and a mermaid-like creature in Papua New Guinea, which natives call the ri, was established as the rare dugong, in 1983.

ranged over a period from 1639 to 1964. Heuvelmans divided the sample into nine categories dictated by characteristics of appearance and behaviour patterns. Some of these he classed as mammals and others as super-eels. The marine saurians, he conjectured, might well be survivors from the Jurassic period.

More recently Professor Paul LeBlond, an oceanographer from the University of British Columbia, has concerned himself with sightings of Caddy. He told Fergus Keeling, 'What it looks like from the descriptions is an animal which is hairy – there are indications of whiskers, of coconut-fibre hair, fairly sharp teeth and eyes which are big and brownish. It could be an unknown representative of a marine mammal related to the seal family, one which lives at much greater depths because they are rarely seen at the surface. Big eyes are most suitable for living at lower light levels in deeper parts of the ocean.'

The body is still missing

If the great wealth of sea monster sightings is truly reflective of a new or surviving species of super fish or marine mammal, it seems reasonable to expect the remains of such a beast to have been caught or washed ashore. Indeed, the sceptics rightly ask why a live specimen has never been captured.

Over the years many strange corpses have been washed up on the world's beaches. Sometimes these seem to exhibit features that correspond with reports of monsters. In 1808 a 55-foot (16.7 metres) long body was found beached on the island of Stronsay in the Orkneys. Before a proper examination could take place violent storms had smashed the corpse to pieces. A drawing showed a long neck, undulating tail and three pairs of legs and fragments of bone provided enough detail for the beast to be identified. Everard Home, a British surgeon who had made a study of shark anatomy, decided that this corpse was a shark. The strange shape, he said, was due to the rapid decomposition of certain bodily parts, namely the lower tail fluke, fins and the lower jaw.

Santa Cruz, California, was the location for a beached 'monster' in 1925. It apparently had a 30-foot (9 metres) neck and a large beaked head. The Museum of the California Academy of Sciences examined the skull and pronounced that it belonged to a rare beaked whale. But scientists do not always agree on a mundane explanation. In July 1960 a large decomposing mass was washed ashore in western Tasmania. Unfortunately, examination of the material did not take place until almost two years later, when scientists from Hobart took away samples in a helicopter. They later released a statement declaring that it was blubber which had probably been torn off a whale. But other marine

A Japanese fishing crew thought they had found the carcass of a sea serpent off the coast of Christchurch, New Zealand in April 1977. Later analysis indicated that the remains belonged to a monster shark

biologists did not agree and thought the explanation unlikely.

The public waited with eager anticipation when pictures were flashed around the world of a huge rotting carcass suspended by ropes on board the Japanese trawler *Zuiyo Maru*. It seemed to show the humped back, neck and head of a water monster. The huge body was netted on 25 April 1977 28 miles (45 km) east of Christchurch, New Zealand. Unfortunately, Captain Akira Tanaka reluctantly tossed the corpse back into the sea once it had been photographed because he was concerned it might contaminate his catch. Study of the pictures convinced most scientists that here too were the remains of a shark. The crew were convinced otherwise, but fibres analysed at Tokyo University were found to contain elastodin – a protein peculiar to sharks. If nothing else, the *Zuiyo Maru* had caught a monster shark. Few grow to a length of 33 feet (10 metres).

Although suspects have been convicted of murder without the evidence of a body, the lack of sea monster remains is a major stumbling block to their acceptance. Professor Paul LeBlond admitted as much in his report on Caddy. 'We will admit that what may pass for sufficient proof in a court of law might not satisfy the criteria of incontrovertible proof. The body is still missing, and the existence of sea monsters may remain in doubt until a specimen is caught dead or alive.'

Remains were found in May 1990 by two scuba divers in underwater caves 32 miles (51 km) off Matagi island, in Fiji. The discovery was made by New Zealander Kevin Deacon and Nigel Douglas, whose family owns Matagi in a trust. The bones of four creatures were discovered in a reef, between 100 and 160 feet (30.5 and 48.7 metres) up a winding passage divided into various compartments. Deacon told reporters, 'We have found what appears to be two adults, one adolescent and a juvenile. They bear no resemblance to any marine creature I know.'

The total length of the adult remains was between 26 and 32 feet (8 and 9.75 metres). Skulls were about 3 feet (1 metre) long. Discovery of the bones was linked to sightings of sea monsters in the region. Fijian legend tells of a huge shark god called Dakuwaqua who could shape-shift and come ashore as a man. Frieda McHugh of Takapuna, New Zealand, drew the media's attention to the memoirs of her grandfather, the Reverend A. J. Small, who recalled first-hand encounters with the beast. While sailing from the Fijian island of Vanua Levu to its neighbour, Taveuni, on a 4-ton cutter in the early 1890s, the boat was shaken by something in the water. Small described it as an immense creature at least 35 feet (10.6 metres) long. When referring to another incident, he wrote:

On another occasion on board a 6-ton cutter travelling to Taveuni it was dusk and the sea was calm. I sat on a deck chair on deck. Later the crew lay on deck and went asleep, also the captain, but he had tied the rope of the tiller around his big toe.

Suddenly I noticed that the cutter was slowly being tugged down. I woke the captain, who exclaimed, 'It is Dakuwaqua', and ordered the crew to throw over their brew of yaquona. Immediately the cutter came up with a jerk, and the shark god swam past. Being dark I could not see it very well, but it was longer than the cutter and had a peculiar shaped head like that of a turtle.

The Dakuwaqua was renowned in native legend for attacking canoes. In those days the beast was tossed a pig, or even a slave to stave off attack. There were other reports recorded by the Reverend Small. In 1912 for instance, he sighted a creature with an enormous dorsal fin, 35 feet (10.6 metres) long and 20 feet (6 metres) wide. It had a short 2-foot (0.6 metres) thick neck, a large turtle-like head and a powerful looking tail. There are also accounts by other witnesses.

Could the remains discovered in 1990 relate to contemporary sightings of sea monsters? A video film of the find was shown to scientists in America but they were unable to identify them. Deacon believes the remains are either prehistoric or contemporary marine animals unknown to science. More photographs and film have been taken and are currently being examined in an attempt to solve the mystery.

THE LOCH NESS MONSTER

The hills slope down over 1000 feet (305 metres) on either side of Loch Ness, the world's most famous freshwater lake. It is one of three lochs which nestle in the Great Glen; a geological rift stretching right across the northern extremity of Britain and formed by a shift in the Earth's crust over 250 million years ago. Loch Ness, the largest and deepest of these lakes, containing 265,000 million cubic feet (7500 million cubic metres) of inky water, is said to be the lair of the infamous 'monster'.

The friendly term 'Nessie' was largely invented by the tourist industry to ensure that the creature attracted visitors. As a result, every year thousands of people from around the world come to Scotland hoping for a glimpse of the elusive and mysterious beast said to lurk beneath the peat-dyed waters.

These sightings pose one of the greatest riddles faced by explorers of the Unknown. Has some remarkable prehistoric animal really survived dramatic climatic and geological changes, or is the unrelenting legend just a complex mixture of imagination, misidentification and blatant fraud?

Sightings of a 'monster' date back centuries. In more recent times folklore references to 'water horses' made by fishermen must be relevant. Soon after a new road was opened on the northern shore bringing more travellers to the loch side, sightings began in earnest. It was in April 1934 when gynaecologist, Robert Wilson, took two photographs from near Invermoriston. Researchers have long believed that these show the neck and head of one of the beasts. The pictures achieved global fame after being carried by the *Daily Mail* newspaper – and the modern legend was born.

Over the years, a catalogue of eyewitness accounts, photographs and underwater explorations have added fuel to the controversy. Barely a summer passes without some scientific, or pseudo-scientific, expedition being mounted determined to solve the mystery.

Betty Gallagher is curator of the Loch Ness Monster Exhibition, a permanent attraction based on the shore at Drumnadrochit. This serves as a national archives for material relating to the phenomenon.

Peter Hough has made many research trips to Loch Ness and interviewed her. Betty told him, 'It is very easy to see things out there. Visitors see monsters left, right and centre. But they're not used to the tricks of the environment. The eye is easily deceived. Normally we receive between three and five sightings a year of what we consider are probably genuine. When a report is made by one of the locals,

The famous 'surgeon's picture' taken in 1934. This more than anything captured the imagination of the public and convinced many of the reality of a monster in Loch Ness. It was not revealed as a hoax until 1994

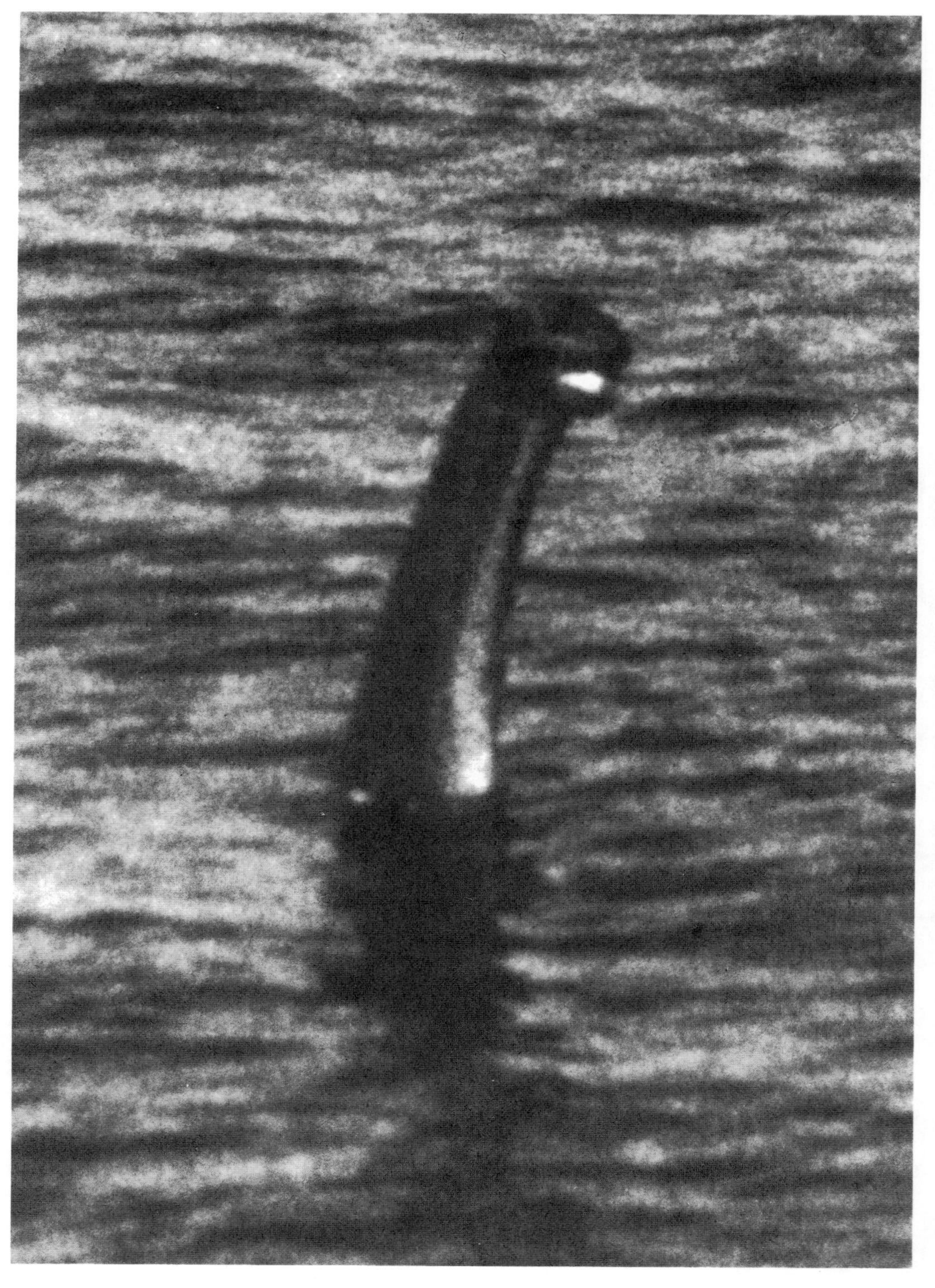

One of two pictures allegedly taken of the Loch Ness Monster by Anthony Shiels. Doubts cast on other Shiels-associated pictures do not bode well for this one either

used to the environment, we think this is it! They know what floating logs and freak waves can do to the eye.'

It is misleading to think of Loch Ness as a 'lake'. A mile (1.6 km) wide, over 23 miles (37 km) long, and with a mean depth twice that of the North Sea, it is more like an inland ocean. Romantic it certainly is. Standing on the shore at Lower Foyers under a full moon, looking out over mill-pond conditions can be a wondrous sight. But the loch can be a 'short-tempered mistress' too. The Great Glen acts as a natural wind tunnel. Peter Hough has been out on the loch in a small boat and has suddenly found himself riding 9 foot (2.7 metres) high waves.

Freak waves can be very mysterious to the uninitiated. One can watch boats pass along the waterway, and ten minutes later see a wave forming out of nothing some distance from the shore. To the casual tourist, already half expecting to see the monster, this could appear to be a series of humps briefly breaking the surface.

Other problems can be created by floating logs, mats of rotting vegetation, creatures such as otters swimming just below the surface and many other natural anomalies. As a result, sightings made by local people, aware of environmental factors, deserve and naturally acquire most attention.

A fisherman's tale

Jimmy Cameron, a greengrocer and fisherman has had several sightings. He told Hough about two of these:

I was out fishing with an old friend of mine – a retired policeman. Earlier an American tourist had heard us talking and asked if he could come along too. We had a 17 foot (5 metres) rowing boat, so it was no problem. We were up at the north end of the loch not far from an old quarry, when we noticed two black humps about half a mile (0.8 km) away. Suddenly they disappeared then resurfaced again. The American took a photograph, and some months later he sent me a copy. But he was really too far away for it to show much. However, I had a much more definite sighting in 1981.

I was driving my greengrocery van heading towards Loch End at about 12.40 in the afternoon. As I came level with a lay-by, the trees were very short and leafless, and this gave me a clear view out over the water. There, about 600 yards (549 metres) away, I saw quite clearly a neck and head sticking out of the water by about 3 feet (1 metre). It was dark brown or black, and the head was flat.

I stopped the van at the next lay-by but it had gone. It had been facing towards Dores on the south shore, but I don't know whether it was stationary or moving in that direction. I could see many miles up the loch towards Urquhart Bay and there were no boats in sight. I'm very sure of what I saw; in fact I sat in the van an hour and a half waiting for the thing to re-appear!

A heavenly witness

Fishermen may, perhaps unjustly, be accused by some of spinning tall tales from time to time, but few would question the word of a man of the cloth. Father Gregory Brusy, Father Prior of the Benedictine Abbey at Fort Augustus, situated at the southern end of the loch, had his own strange experience in 1973. He explained it as follows:

It was a lovely October morning. I had the organist of Westminster Cathedral staying with me, and we decided to go for a short walk. We stood on the stone jetty looking out across the bay. The water was calm and there was not a boat in sight. Suddenly, about 300 yards (274 metres) off shore, there was a great churning of the water – more commotion than a fish could make – then a 5-foot (1.5 metres) length of neck appeared with spray flying everywhere. We observed it for twenty seconds as it began to move away before diving back beneath the water in a sort of curious sideways motion.

Many who have seen and sometimes photographed an anomaly in the loch have no doubts that they have witnessed a real animate creature of unknown origin. If this is so, then what could it be?

The majority of sightings describe part of a back, a long neck and flat head. This leads most experts to speculate that there might be a colony of plesiosaur-type creatures which were trapped in the loch 7,500 years ago, when it became cut off from the sea in the wake of local volcanic activity.

The plesiosaur was a sea-faring giant which geologists believe died out with the other dinosaurs millions of years before pre-hominid man evolved. Yet there are other forms of life just as ancient and which were also thought extinct for millenia. The most famous example is the coelacanth, a huge fish which had been pronounced extinct but was found to be alive and well in the Indian Ocean in 1938. It was almost unchanged from fossilized examples 70 million years old.

Frank Searle, a former monster hunter, and controversial figure, has his own theories. He spent fifteen years living at the loch side, and took many pictures. Several of them have been condemned as hoaxes. In particular, one showing a head and neck, was compared with an illustration of a brontosaurus published as a post card. Even so, his enthusiasm and common sense when discussing Loch Ness cannot be faulted.

If we are dealing with surviving families of fish-eating plesiosaurs zoologists argue that they probably used Loch Ness as a breeding ground when the loch was originally open to the sea. As it slowly changed from saltwater to freshwater, they must have evolved with it. The sightings which are considered genuine usually last just a few seconds – a commotion in the water, a neck and part of a back bursts into the light, then it is gone again. If the creatures are purely marine dwelling, like fish, they would not need to surface. When they do it is probably by accident.

Imagine a 30 foot (9 metre) beast coming up from the depths as it chases the salmon which swim just a few feet below the surface. Sometimes, perhaps, it cannot stop its momentum and breaks the surface momentarily, before sinking again below the black waters.

A supernatural beastie?

Some researchers suggest that there might be a paranormal explanation for Nessie. One idea is that the geological fault lines in local rocks trigger electrical activity in the atmosphere. Dr Michael Persinger, the Canadian neurophysiologist, has speculated that this energy could affect the brains of individuals who advance too close, leading to powerful hallucinations. Is Nessie a vision conjured up by natural energy forces? If so, why should everyone have the same vision and how could these be filmed?

Another theory is that the creatures are some sort of 'video recording' from the past encoded into the environment. When certain conditions prevail this may 'replay' as a living, holographic scene from out of the Jurassic or Cretaceous era. But how can waves strong enough to rock passing boats be created by a visionary phenomenon?

Naturalist Adrian Shine who has been leading monster-hunting expeditions at Loch Ness and Loch Morar since the 1970s. Shine has no truck with those who believe there might be a paranormal answer to lake monster sightings. He believes the beast might turn out to be a giant fish

There have been many attempts to obtain fresh evidence on Loch Ness. Here, the airship *Europa* drifts above Urquhart Castle watching for anything unusual. The Loch Ness Project research barge can be seen on the water

Anne Arnold Silk, a British researcher, points the finger at the active fault which runs right through the Great Glen. She speculates that standing wave formations created by seismic stress could create the illusion of something large moving in the water. Most scientists are fairly scathing of such imaginative suppositions.

Adrian Shine is a naturalist who has spent many years at Loch Ness carrying out observations to determine just what exactly might be the source of all these reports. He says, 'I don't know anything about the paranormal. I'm a naturalist. Possibly we're talking about a large fish. The Atlantic sturgeon for example can grow up to 20-foot (6 metres) long. There is something in the loch of zoological interest, we've known that for years. We've also known that the alleged evidence is faulty. But there is no scientific reason why there shouldn't be anything in Loch Ness.'

I was that monster!

A new breed of sceptics argue strongly that not only is the evidence faulty, but under close scrutiny, it is non-existent. Their case was strengthened in 1994 when the truth behind the famous 'surgeon's picture' was uncovered by two researchers with the Loch Ness and Morar Scientific Project, David Martin, a zoologist, and colleague Alastair Boyd. The trail started with a small diary story in *The Sunday Telegraph* published in 1975. An Ian Wetherell claimed that he and his father had once hoaxed a photograph of the Loch Ness monster. Ian was the son of the late Duke Wetherell, a flamboyant film maker who once faked some monster footprints on the shore near Fort Augustus.

Eighteen years later Martin and Boyd tracked down an elderly man called Christian Spurling, Duke Wetherell's stepson, who after sixty years gave details of the fraud.

It emerged that after the footprints had been

revealed as frauds, Wetherell returned to London feeling angry and wanting revenge. He sent Ian out to Woolworths to buy a clockwork toy submarine and several tins of plastic wood. Spurling, a keen model maker, agreed to fashion a neck and head onto the submarine's conning tower. It was then floated in a quiet bay at Loch Ness and the photographs were taken. Now all they needed was someone with credibility to take the credit.

Thinking it was just a harmless bit of fun, gynaecologist Robert Wilson agreed to be the 'photographer' and was given four photographic plates to take into an Inverness chemist shop. The rest is history.

Elementary, my dear Nessie!

Possibly the most extraordinary evidence generated in support of the existence of the Loch

When these boats began scouring the loch in July 1987, they reported mysterious movements in the water

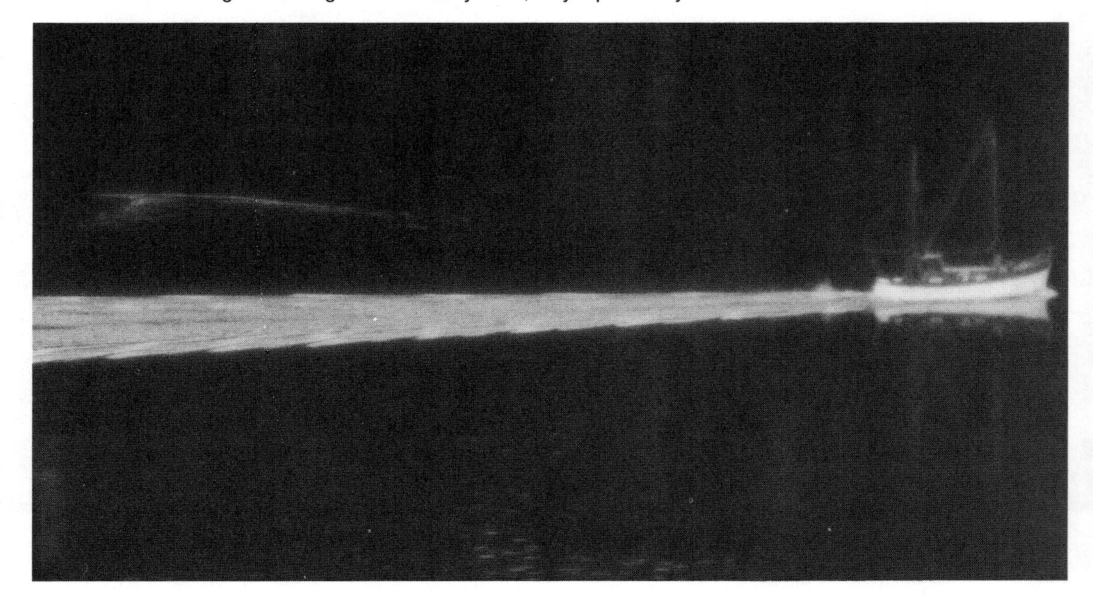

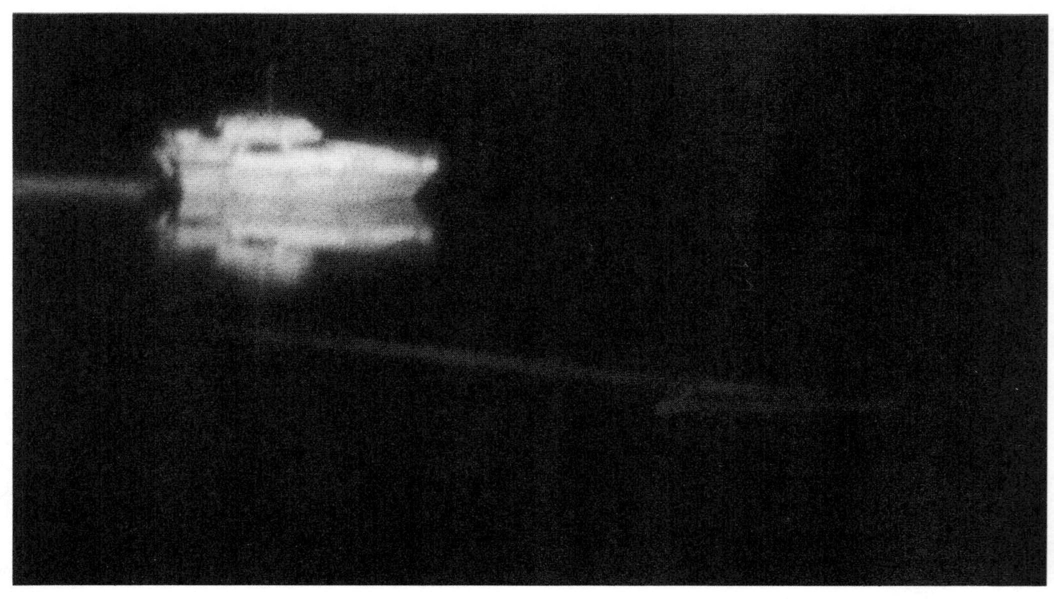

Operation Deepscan was the most ambitious project to date to probe the loch's murky waters. In October 1987, twenty-two echo sounders were strung out between a line of boats as they swept Loch Ness. Some interesting results were obtained

Ness beast are underwater pictures taken by an American survey vessel. These made international headlines at the time, but great controversy now surrounds them.

In 1972, Dr Robert Rines representing an organization called the Boston Academy of Applied Science produced two underwater photographs taken in Urquhart Bay which appeared to show the diamond-shaped flipper of an unknown underwater creature. At the same time, according to Rines and his associates, a large object was tracked on sonar. These pictures were indeed impressive. But what was not generally understood was that the originals showed nothing more than fuzzy blobs. The images released to the public were computer-enhanced versions 'cleaned up' to take account of the impenetrable murk of the Loch Ness waters.

Three years later Rines came up with some more underwater shots, this time showing a 'gargoyle-shaped head' and 'upper torso, neck and head' of the 'monster'. Sadly, this is now widely believed by disappointed researchers to be a giant plastic model monster which sunk in Urquhart Bay during filming of the film in 1969 *The Private Lives Of Sherlock Holmes.*

The 'gargoyle head' may also have a prosaic answer. In October 1987 divers, searching in the spot where Rines took his picture, recovered a rotting tree stump, which when viewed at the correct angle bears more than a passing resemblance to the famous 'head of Nessie' photo.

News from the loch side

Nicholas Witchell, the BBC newsreader, is no stranger to the Loch Ness scene. His book *The Loch Ness Story*, originally published in 1975, has been extensively re-written as the original had leant heavily on Rines' underwater 'evidence'. At that time, Witchell also favoured the plesiosaur theory. Peter Hough asked him how he felt about it all now, having had some years to reflect upon the evidence.

Authorities and sources

■ *Steuart Campbell* has produced the most sceptical research of all. He is an architect and maverick of the paranormal who has fiercely disputed the evidence for many phenomena – especially that which has been captured on film. Although his strident views sometimes antagonize people, his research is at times first class and some impressive evidence has crumbled under his scrutiny. He believes Nessie is a myth forged from misperception and wishful thinking. However, his 'rational' explanations in the field have at times stretched credulity to breaking point. His only book to date is *The Loch Ness Monster: The Evidence* (Aquarian, 1986)

■ The late *Tim Dinsdale* spent more time researching and writing about Nessie than almost anyone else. He produced his own filmed evidence of an object speeding through the waters off Lower Foyers, although some contend it shows a motor boat. He has written several books on his work: notably *The Leviathans* (Routledge and Kegan Paul 1966, updated by Futura in 1976) and *Project Water Horse* (Routledge & Kegan Paul, 1975).

■ *Roy Mackal* is a zoologist from Chicago who has mounted expeditions all over the world in search of legendary animals. These include several to the Belgian Congo seeking a dinosaur very much like Nessie reported by generations of natives. His research into Loch Ness was published in the well-argued book *The Monsters of Loch Ness* (MacDonald 1976).

■ *Michael Meurger*, in his book *Lake Monster Traditions* written with Claude Gagnon, conducted a cross-cultural analysis. He concluded that cryptozoologists and the sceptics had both got it wrong. '. . . the moment of the experience', he wrote, 'is less important than the later re-working of the sighting, and the social pressure upon it, to fit cultural patterning.'

■ *Nicholas Witchell* has brought a professional newsman's status to the field. His book *The Loch Ness Story* (Dalton, 1974) was updated by Corgi in 1982 and more recently still in 1990.

I think the Rines pictures were an error which was made in 1975. They should no longer be considered as valid evidence. I now have no views on what might be in Loch Ness. I just see myself as a journalist recording the experiences and opinions of others.

The new book includes a section written by Dr Denys Tucker. He was a principal scientific officer at the British Museum of Natural History until 1960 when he was sacked for some of his unconventional views. He believes there are creatures in Loch Ness, similar to the plesiosaur.

Witchell was at the loch at the time when the rotting tree stump was hauled to the surface, so it is not surprising that he has drastically reassessed his opinions on the Rines' evidence.

Sweeping the loch

Over the years there have been some wonderful gimmicks used during serious scientific expeditions: probing the waters with a yellow submarine, and even using a supposed female-monster mating call to attract Nessie to the surface. But one of the most ambitious projects was Operation Deepscan.

Adrian Shine was the project leader for a massive sonar sweep of the entire width of the loch sponsored by a long list of companies including Lowrance Electronics of Tulsa, Oklahoma, Swiftech Ltd of Wallingford, England, and the Scottish Highlands and Development Board. During its operation between 8 and 12 October 1987, Deepscan did come up with some interesting results, if well short of media hopes. Adrian Shine explained:

Deepscan was a more ambitious follow-up to our earlier trials held in 1982. On that occasion we had used scanning sonars, but there is a strong risk of interference with these. This time we used twenty-two short operation multiple X-16 echo sounders strung out in lines along two parallel rows of boats. These operate vertically downwards through temperature layers, and won't make inanimate objects such as bits of old fishing nets, and rocks, appear to be animate.

Some of the soundings we recorded were later followed up with underwater television cameras, and we found all sorts of debris, including that interesting tree stump in Urquhart Bay! But altogether there were three contacts made in mid-water which we couldn't explain. One of these was a very strong echo which was not there when the second line of boats passed over.

A hilarious rendition of Ogopogo at Kelowna, British Columbia. The sign at Okanagan Lake makes it clear that the monster has a historical background, and is not an invention of modern times

Nicholas Witchell added a media perspective, 'Deepscan was never going to live up to the expectations of the media. The boats used in the exercise could only survey sixty per cent of the total water surface – and the sonar sets had to be turned right down to avoid interference. Even though three curious contacts were made, on the Sunday, this was the day the press corps were preparing for home. Earlier there had been nothing, so the media reported the whole thing as a flop.'

Nessie's cousins

There are reports of lake monsters from all over the world. Even Scotland boasts another: 'Morag' the beast of Loch Morar. After Nessie, probably the most famous is Ogopogo, a resident of Lake Okanagan in British Columbia. This is not an invention of the tourist industry. 'Ogopogo' was sighted by native indians centuries before the white man set foot on Canadian soil in 1860. Another Canadian monster is Manipogo sighted in Lake Manitoba.

In 1609, the French explorer Samuel de Champlain, saw a serpent-like creature in an area of water straddling the border with America. His name was given to Lake Champlain, and the beast, which is frequently recorded, became Champ. The lake is 105 miles (169 km) long and 12½ miles (20 km) wide – ample space to hide a species of unknown creature.

Загадка озера

С июня по октябрь этого года мы работали в Янском районе Якутской АССР. Мы исследовали четвертичные отложения Куларского хребта и прилежащих районов. Состав экспеди-

чество рыбы: чир, форель, щука, баранатка, и только в таинственном озере, как уверяют, нет рыбы. Не садятся на его поверхность ни гуси, ни утки. Озеро «термальное» — оно за-

Below Sketch by Russian biologist, N.F. Glodkikh, of the water beast he saw in Lake Khaiyr, Siberia in 1964. *Left* Another comic model of a monster. This time of 'Issie', a serpent which was recently filmed in Lake Ikeda, Japan

Every continent has its lake monsters. Lake Vorota, Labynkyr and Khaiyr in Siberia have generated many sightings over the years, some by responsible scientists. Swedish lakes too have been the location for encounters with water beasts – the most famous being Lake Storsjo. The Irish lakes have also earned a place in monster history – despite the fact that many of them seem too small and shallow.

The first film of a Japanese monster was taken on 4 January 1991 at Lake Ikeda. Two long dark shapes were captured on nine minutes of video film by Hideaki Tomiyasu and his family. The 30-foot (9 metres) object repeatedly dived and resurfaced until a motor boat seemed to scare it off.

1992 – The story continues

The summer of 1992 was a vintage one for Nessie. After some years of shy retreat she surfaced regularly to recapture the public's imagination.

In July Project Urquhart attempted a new ambitious sonar sweep of the waters, playing down publicity claims that they were monster hunting. The scientists used *Simrad*, a sophisticated Norwegian vessel, and claimed that the three-week expedition, the first real survey of marine life in the loch, was that and nothing more. But at 19.04 hours on the evening of 27 July something 'large and solid sounding' was tracked by the sonar beneath the waters between Foyers and Invermoriston. This was the largest sonar reading ever made in the middle of Loch Ness, baffling the experts.

The team refused to speculate on what they had detected, and are equally bemused by a mysterious line of dots 600 feet (183 metres) below the surface, running along the bed of the loch approximately 200 feet (61 metres) apart.

They appear so regular that they must be the product of an intelligence.

No sooner had the fuss died than a 45-year-old man camping at Fort Augustus had a sighting on 9 August. While brushing his teeth in the loch at 07.00 hours he spotted a brownish black neck estimated at 6 feet (1.8 metres) in length, with a flat head which seemed to stare at him just 100 feet (30.5 metres) off shore. The man grabbed his pocket camera and took four colour shots before it slid under the waters. Some minutes later, further out, four humps appeared to break surface in the loch. Rivulets of water ran off the protuberances indicating this was not a freak wave.

The pictures do show something odd but photographic analysis by the Royal Air Force and Kodak struck a cautious note. There was no evidence of tampering with the film, photographic experts stated, and the pictures do show something.

Six days later, for the first time, a camcorder filmed Nessie from a position above Urquhart Castle. Something black and apparently solid can be seen causing a commotion in the water some distance from the shore. Adrian Shine described it as a 'cycling motion', but speculated it might be a freak wave, an option which Steuart Campbell cites as the explanation. However, zoologist Peter Meadows regarded it as more intriguing. He told reporters, 'I'm amazed by what I've seen, I've never seen anything like it before!'

Local legend says that the loch never gives up its dead. Will the truth about Nessie likewise remain hidden in the depths forever?

Nessie: the facts

■ First known sighting was recorded in AD 665 when the Abbot of the Isle of Iona produced a biography of St Columba. This told how the saint had driven a water monster out of the River Ness near Inverness.

■ First modern recorded sighting was made in October 1871. Then, a shape like an upturned boat was seen churning up the water below Abriachan.

■ Strangest early report occurred in 1880 when there were several claims by children of a large land animal the colour and size of an elephant waddling down the slopes and into the lake. There have been other land sightings but none in recent times.

■ The first known photograph was taken on 12 November 1933 by Hugh Gray. It shows a snake-like object writhing in the water near Foyers. Some people now think this is an out-of-focus image of a dog swimming towards the camera.

■ First moving picture was taken in 1935 by a retired doctor. Four minutes of allegedly clear film show Nessie floating on the surface in all her glory. This was put in a London bank vault to remain there until people took the subject seriously. It is reputedly still there with nobody knowing who owns the right to release it.

LIVING DINOSAURS

Forgotten rulers of earth

Two hundred million years ago the earth was a sub-tropical hothouse filled with strange and frightening creatures. There were no human beings. Only a few mammals reared their heads then – mostly small shrew-like creatures that found themselves easy prey. They were vastly outnumbered by the thousands of different varieties of dinosaurs.

These beings filled the air and the seas. They roamed the forests and the deserts. Between them they completely dominated every other form of life for tens of millions of years. In contrast we have been here a very short time.

Suddenly and quite mysteriously, however, their domination ceased. In what was geologically speaking the blink of an eye millions, probably billions, of these beasts died, leaving only their fossil remains to show that they had ever been here. Just why did the dinosaurs disappear so dramatically? If it happened to them could it also happen to their successors, *Homo sapiens*? Are we but transitory rulers of a fragile world? Did the dinosaurs really become extinct or

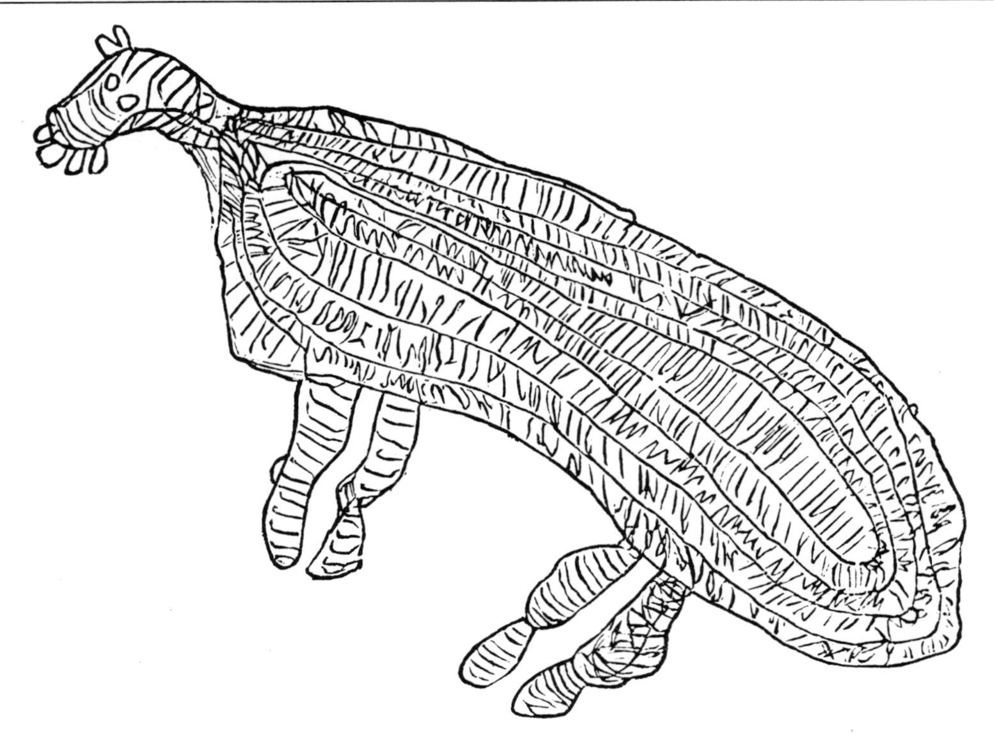

Monster legends illustrate the longevity of sightings. This Australian aboriginal drawing of a bunyip was discovered in the Murray River area in 1848. Are there really places left on this planet for dinosaur descendants to hide, or must we seek even more bizarre answers?

somewhere in the furthest, darkest corners of this planet do survivors of that once magnificent race still exist? There is a surprising wealth of evidence to suggest this might be so.

A menagerie of monsters

Remarkably, in this age of satellites and computers, there are still huge tracts of land that are relatively unexplored. We also know next to nothing about the deep seas that fill two-thirds of the surface of our planet Earth. If witnesses are to be believed there are still dinosaurs living over large areas from the densest jungle to the skies above the most arid desert.

In the African Congo searches have been made to find the *mkoele mbembe* – which by all accounts is nothing less than a creature that is akin to the *brontosaurus*, said to have last walked the marshlands of the Earth tens of millions of years ago.

Above the hot desert lands of south-western USA witnesses see something very different – a flying dinosaur with jagged teeth and folded wings. It goes by a name that means 'terrible flying lizard'.

A world away, the island continent of Australasia is not free from such tales. Native aborigines have long told of gigantic creatures that roam the outback – today stylized in legend as the terrifying bunyip.

Just how can these monsters survive without one being placed upon display or paraded before television cameras in the way of the fictional 'King Kong' and 'Godzilla'? There are two ways to approach this question. The sceptical scientists say that the solution is obvious. If they did exist they would be caught, but as they have never been caught they clearly do not exist. This brand of circular reasoning seems superficially persuasive, but it does have its drawbacks. It still has to account for the sightings that have been made of these animals. It does so by arguing that they are all misperceptions of ordinary, known creatures, such as extra-large fish or big lizards.

The hunters who believe that this is the wrong approach do include scientists. They have their own branch, named cryptozoology (the study of unknown animals) and accept the testimony of witnesses. They have mounted some remarkable expeditions to try to find these creatures. These expeditions have resulted in tantalizing but as yet not fully convincing results. However, most cryptozoologists seem convinced that it is only a matter of time. Somewhere out there these animals do survive and one day indisputable proof will be secured.

In answer to the sceptics the cryptozoologists point quite clearly to other surviving species once thought to be extinct but now known to be very much alive.

The survivors

During the early 1800s travellers returning from the islands of the Pacific had some incredible tales to offer. They spoke of a bizarre creature three times the height of a man which looked like an ostrich and was apparently a bird but one that could not fly. People scoffed. It was a ridiculous story. If such a creature existed then science would know all about it. But science did not know about it. Yet it did exist.

The creature was a moa. It was very rare and the Western settlers of the islands soon despatched the final few to the geological record books. You can see stuffed examples or models of them in major museums. Even today there are stories that moa have been sighted and that perhaps not all were killed in the fear-soaked purges of 150 years ago.

Of course, a remote island might well have sported an exotic life form in the nineteenth century. We might even just about accept that a few could survive today and escape capture.

An 1861 illustration of hunters stalking a moa, a giant flightless bird now thought to be extinct, although a few may still survive in very remote island locations

A rare photograph of the thylacine, or Tasmanian Tiger, now believed extinct in its only habitat of Australasia. But sightings of the 'Tassy Tiger' continue to be reported in the popular media

A mystery animal – akin to a thylacine – photographed in Victoria, Australia, in 1964. Its identity has never been established. Many unknown creatures are said to exist within the vast bush of Australia

These are fairly remote places where civilization barely treads. The same could surely not be held true for an inhabited land with a modern culture?

Australia has developed alone throughout much of geological history and cut off from other land masses. It offers some of the strangest and most unique wildlife in existence. Interestingly, the marsupials, or pouched animals, familiar to us in the form of kangaroos and wallabies, all show a kind of parallel evolution with species on other continents. Although they are different they fill the same gaps in the hierarchy of creatures found elsewhere and often resemble pouched versions of these same animals.

However, there is a gap is this neat order. Nothing akin to the big cats seems to exist. Until the early twentieth century there was the Tasmanian Tiger (or thylacine): this is not a marsupial big cat but a striped animal that looks rather cat-like and yet is related to a wolf. The 'Tassy Tiger', as locals call it, 'survives' in the occasional newspaper story and hunters' expeditions. One was reliably observed for several minutes in 1982 at very close quarters and a somewhat controversial photograph of one was taken on the mainland in Western Australia at about the same time.

There are also reports of a real tiger-like marsupial. These have been made with such consistency that zoologists, once sceptical, are now looking at the evidence anew. In 1970 serious attempts were made to hunt the bushland around Cairns in far North Queensland to find the so-called Queensland Tiger. Some fifty-nine sightings were made but no hard evidence uncovered. The animal thus remains in doubt but as it fills a gap in the zoological record its existence, rare as it must be, remains a distinct possibility. Unlike living dinosaurs, scientists find it easy to accept the tigers' existence.

Australia still has thousands of miles of uninhabited wilderness. This is definitely not true of England, one of the most densely peopled lands on Earth. Yet it is now widely accepted that at least two colonies of wallabies live in Britain. Wallabies are certainly not native to the northern hemisphere. It seems that they have thrived from being escaped pets or zoo animals.

Although the wallabies' existence is well attested most people in Britain do not know about them. They are only rarely seen and even more rarely photographed. Yet one of their prime habitats is the Derbyshire peaks around Buxton only 20 miles (32 km) from the major cities of Manchester and Sheffield and in a place where during the summer countless fell walkers and picnickers visit.

Britain also has sufficient evidence, from sightings and photographs, to prove to reasonable satisfaction that a few big cats (akin to the puma or lynx) live in some wild areas such as Dartmoor. These have definitely not been native to the islands for thousands of years, so it is thought that escaped pets have founded colonies.

We are here talking about creatures that we know are real or became extinct just a short time ago. This is very different from the suggestion that dinosaurs live on. Officially, the last dinosaur became a fossil over 60 million years ago. If any dinosaur had survived past that era there should be indisputable evidence within the chain of fossil records, let alone clear sightings or scientific proof.

A rare photograph of Britain's own mystery animal – the phantom 'Big Cat'. In August 1966, during a spate of 'puma' sightings in Surrey, a police officer photographed this unidentified big cat at Worplesdon

Scottish big cats have been hunted for many years but rarely has any hard evidence surfaced to match the many sightings. This unidentified black wild cat species was captured at Kellas in 1983

In nature fossilization is very much the exception. For an animal skeleton to be preserved in rock, a chain of freak events must take place. This occurs very infrequently. Many fossils have been discovered because countless creatures have died during hundreds of millions of years enabling freak events to take place many times. If, however, only a handful of survivors from a species is left, then their deaths, even over many years, would be far smaller in total and few would by chance become fossilized.

The coelacanth provides the proof that this is possible and not just idle speculation. This is a large and ugly-looking lung fish, a distant cousin of the marine dinosaurs. Scientists were agreed that it died out 70 million years ago. Their reasoning was impeccable: it was several feet long so if it was still in the oceans we could hardly have missed discovering it in modern times. But, more importantly perhaps, it had vanished from the fossil record that far back. If it had lived on into the present then there would have to be fossil evidence. But in 1938 a living coelacanth was fished up from the Indian Ocean off the coast of Africa. Since then it has been discovered that this 'living fossil' is quite abundant. The fossil record has misled us and not the fishermen's tales which for so long described sightings of a 'monster' that scientists had refused to believe existed.

If a 70-million-year-old fish can be found in the twentieth century, the cryptozoologists contend that nothing is impossible. It is no major step towards the idea that a few even more remarkable species might still be out there in some isolated or inaccessible spot.

The flying nightmare

Most people think of pterodactyls as huge flying dinosaurs with big wings and jagged teeth. In fact they were not dinosaurs at all. They ranged in size from the tiniest at a few inches across to the 18 feet (6 metres) wingspan of the *pteranodon.* Although *quetzalcoathus* has a wingspan of 40 feet (12 metres) and was found in the south-west of the USA. In 1972 a near-complete example of *pteranodon* was found fossilized in rocks in Texas. The largest living bird today has wings spanning only half the

A fossil pterodactyl – flying lizards which became extinct millions of years ago. Witnesses still claim to see them

width of the *pteranodon.* Today surviving pterosaurs are said to exist and, more remarkably still, some of the most frequented locations are the desert-like canyon lands of Texas where the rocks provide some of the finest fossilized examples of these same creatures from the height of their domination of the skies.

In mid-February 1976 two sisters claim to have seen a huge bird with teeth, the face of a bat and black leathery wings in the Big Bend National Park of Texas. They were later shown sketches and immediately identified it as a *pteranodon.* On 24 February that same year one of the best sightings of all occurred near San Antonio. Three teachers saw a 20-foot (6 metre) version fly right over their car. Patricia Bryant said of it, 'I could see the skeleton . . . it stood out black against the background of the grey feathers. It had previously been thought that pterosaurs were featherless.

Another witness, David Rendon, observed that the wings were comprised of many small features forming a membrane and looked very boney. He also reported how it soared overhead, gliding on currents of air. Until very recently palaeontologists thought that pterosaurs glided like bats and were incapable of real flight. But in 1988 the discovery of an unusually well-preserved *pteranodon* fossil challenged this. An entire network of tiny bones was found in the wing section and traces of feathers were also evident. It had been assumed (given that soft tissue is almost never fossilized) that a leathery, featherless skin would have covered the pterosaurs. Now scientists are not so sure. However, the discoveries which led to this

The goanna, just one of the large lizards found in Australasia. But much larger, dinosaur-like, creatures are reported from remote bush areas

change of heart were made twelve years after supposedly living *pteranodons* had been seen drifting above the Texas canyons – *pteranodons* that were displaying the seemingly correct feathers and intricate bone structure now thought to have applied.

Does this mean that these sightings were authentic? If so, how can such gigantic creatures have survived in the state of Texas for so very long and yet so seldom be seen and never photographed?

The amazing Australian lizards

Australia is full of lizard-like creatures, from the friendly goanna to the fearsome-looking beasts that might be descendants of dragons. These are large monitors – tropical lizards. In the uranium-rich Arnhem Land far to the north some fine examples can catch the visitor unawares.

Jenny Randles was lucky enough to be invited into this huge aboriginal country where whites have free access only one day a year. The wildlife here is spectacular and she heard stories of giant lizards that had scared young children bathing by a dangerous river. She spoke with a ranger in the adjacent splendour of the Kakadu National Park to the east of Darwin, where the *Crocodile Dundee* movies were filmed. He told her that many legends abound about huge bunyip or dragon monsters which are oversized monitor lizards and occasional land-faring crocodiles. He had himself seen one almost 20-feet (6 metres) long saunter across his path while in the bush one day.

Are these simply large but mundane creatures who live in the relatively predator-free wild bush? Or is there more to these tales which date back millenia and tell of dinosaur-like monsters that terrified the local aborigines?

Rex Gilroy conducted specialist research into the mystery animals of Australasia while he was curator of the Natural History Museum in Mount York, New South Wales. He found that apart from the bunyip, said to reach 30 feet (9 metres) in length, there was an even more feared beast in aboriginal tradition, that of the 50-foot (15 metres) *mungoon-galli*.

One report he collected came from a farmer in the Wallangambie Wilderness where he allegedly came upon a giant reptilian beast eating one of his stray cattle. He kept his distance but was impressed by the size. On 27 December 1975 a farmer at Cessnook saw the beast and described it as 30 feet (9 metres) long and with mottled-grey skin sporting dark stripes. It was raised 3 feet (1 metre) off the ground on four thick legs and had a head and a neck that was 3 feet (1 metre) long and 2 feet (0.6 metres) thick.

A terrifying account comes from Frank Gordon, a scientist researching snakes in the Wattagan Mountains amid the vast hinterlands. He was sitting in his Land Rover one day in 1979 recovering from several hours fruitless skink (water-snake) hunting. Beside him was a large log. Suddenly the log rose up into the air and revealed itself as a well-camouflaged lizard monster far larger than his truck. After he regained his composure he estimated that it must have been almost 30 feet (9 metres) long.

Gilroy has come across several similar stories of travellers on the dusty outback trails who stop to move a large felled tree covering the entire width of the road only to discover that it is the body of a giant lizard monster. He is convinced that these cannot be ordinary lizards which are never longer than 12 feet (3.5 metres) at the most in Australia. He believes they might be related to the fearsome frilled lizard, the Komodo Dragons, found just across the Indian Ocean in Indonesia and said by many to be the closest living creatures in appearance to the dinosaur. Indeed such animals are often filmed for 'dinosaur' movies and enlarged by way of special effects.

The Komodos reach a mere 9 feet (2.7 metres) in length, so whatever creatures are inhabiting the Australian bush are much more impressive than they are. Fossils show that giant monitors of over 20 feet (6 metres) were around in the not too distant prehistoric past. Indeed, the fossil records show that they first appeared around 60 million years ago, exactly when the dinosaurs became extinct. These may indeed have triggered aboriginal legends. Rex Gilroy hopes that one day definite proof of these spectacular creatures will be secured.

Operation Dinosaur

The African Congo has attracted a great deal of attention in recent times with stories that a dinosaur colony exists in the remote swamplands. These stories have persisted since the earliest days of exploration.

In 1776 Abbé Proyart wrote about the first missionaries to reach the Congo. They discovered tracks with prints almost 3 feet

The fearsome-looking Komodo dragon, found on Indonesian islands

(1 metre) in diameter, suggesting a monster at least 25 feet (7.6 metres) in size. During the 1800s early traders were shown drawings on cave walls made by the natives which depicted the creature who allegedly made these tracks. They said that it was rare but did still exist.

The most favoured location was the Likouala, first visited and explored in detail in 1913/1914 by Captain Freiherr von Stein. He took pains to collect stories from native guides about the so-called *mokele-mbembe.* The animal was said to live in small pools or adjacent caves. It was strictly vegetarian and ate a variety of riverside liana plant called the molombo. Eyewitnesses told how the monster was the size of an elephant and had 'smooth skin, brownish grey in colour . . . a long flexible neck and a single [very long] tooth or horn.'

Reports alleged that natives had been killed by the animal, but it was not thought to be malicious or carnivorous. Instead the deaths normally resulted from canoes overturned as the creature surfaced from a lake and its huge tail swamped or sank them. The von Stein expedition saw no such animal but did find tracks and was persuaded by the consistent stories. These continued to be collected during the following years when occasional scientific surveys reached the area.

In February 1980 a mission, led by cryptozoologist, Dr Roy Mackal from the biology department at the University of Chicago, went to Likouala. Mackal and his team were specifically looking for dinosaurs. In the villages around Epena, Mackal talked to eyewitnesses. The creature was said to be up to 40 feet (12 metres) long in some cases. He heard how the few that remained were in the Lake Tele area and in 1959 a gang of pygmies had seen two disturbing their fishing as the animals moved from river to lake and back again. The bushmen had constructed a barrier, trapped one beast and after killing it with spears cut the carcass for a mammoth feeding session. The task was said to have taken a long time because of the long sinewy neck of the creature.

After the expedition Professor Mackal said that he was convinced the *mokele mbembe* was real, that very few may now survive and it was important to try to secure evidence or a specimen before they all died. In October 1981 he returned with more researchers, including Richard Greenwell. It was proving very difficult given the political situation to get permission to enter the area and not all the natives were friendly. Some were reputedly cannibalistic. It would be impossible to make the enormous trek by boat and on foot without close local cooperation. Other hazards ranged from big cats to crocodiles and killer bees.

Indeed, the expedition did set off but was unable to reach Lake Tele, where the monsters were said to be, because the guides decided this four-day leg of the journey was too dangerous.

However, they almost succeeded in their task they thought, hearing huge splashes of water and finding inexplicable gaps smashed through the bush. The expedition proved that the region was geologically identical to the Cretaceous period – a period when herbivorous dinosaurs most resembling tales of the *mokele mbembe* had inhabited the area.

An American adventurer, Herman Regusters, set off on his own trip at the time, persuading natives to guide him, his wife and several scientists from the Congo into Lake Tele. Here, on 29 October 1981, according to his report, 'every member of the scientific team except the photographer' saw the long neck of the dinosaur surface and then the creature swimming about the lake for several minutes. Regusters told the world that this creature was without doubt a dinosaur. It was like a *brontosaurus*, the mild-mannered long-necked creature popular in monster movies, but much smaller. He suggested it was more akin to an evolved version of the *atlantosaurus*. The fossil records suggest that this was probably the last surviving saurian in the Congo area which would support this suggestion. But the mystery remains: had the world finally found its living dinosaur?

The search for monsters

After his return Dr Mackal struggled to obtain more funding and to beat the African political system. However, the Congolese themselves sent a team into Likouala in early 1983. It was headed by Dr Marcellin Agnagna from the Zoological Gardens at Brazzaville. He had joined Mackal on the 1981 expedition so was the ideal choice.

Agnagna's aim was to come back with documented proof of the creature's existence. Unfortunately the outcome of the expedition was one of the greatest controversies of cryptozoology. During the first couple of days of May 1983 a native was washing in the lake when the 'dinosaur' surfaced. Agnagna was some yards away filming monkeys. There was only a little film left in the camera, but he responded to the yells and rushed towards the shore. There it was - the *mokele mbembe* 500 feet (152 metres) or so away, in full view. Agnagna was convinced he was capturing a historic event and ignored the natives cries to run and instead walked towards the creature. He waded into the water's edge pointing the camera straight at the beast, letting the final few minutes of film wind away.

The creature was reddish brown. The neck stretched like a giraffe's so that it could pluck the molombo from the tall waterside trees. Agnagna got close enough to see that the eyes were like those of a crocodile but no teeth were visible. He estimated, from the parts exposed above the water, that its size was smaller than had been thought previously – perhaps 15 feet (4.5 metres) or so. It was undoubtedly a living dinosaur. As he waded into the pool the water became too deep so he had to stop, watching in quiet astonishment as the dinosaur continued to bask in the sun ignoring its human watchers. After twenty minutes of what must have been the strangest close encounter in history the animal seemed to grow bored and sank back beneath the surface of the lake.

On his return home, Agnagna was adamant he had seen a living dinosaur telling the world, 'It can be said with certainty that the animal we saw was *mokele mbembe*, that it was quite alive . . . [and] is a species of sauropod living in the Likouala swamps.'

Sadly the film was a blurred disaster. Agnagna had been filming on macro setting in order to gain close ups of the monkeys and forgot to reset the camera when wading out after the dinosaur. Several people had taken shots from the shore, but only one film survived the heat and humidity of the several weeks of trekking back to Brazzaville. When the film was processed it came out blank. It seemed that somebody had left the lens cap on in the excitement.

Roy Mackal told Jenny Randles when they met at a conference in Nebraska in November 1983 that Agnagna had expressed great regret at this terrible disappointment, saying, 'It may be hard to convince people because I didn't get pictures.' But he pledged to return.

Jenny shared this sad story with a friend, Bill Gibbons, upon her return to Britain. He is an adventurer, ex-army survival expert and paranormal researcher. Their conversation was to have profound consequences.

Dinosaur hunting for Queen and country

Soon after Jenny Randles told Bill Gibbons of Mackal's stories about the *mokele mbembe* a friend of his came into a large sum of money. Bill was already intrigued by cryptozoology and persuaded his companion to devote the sum to a great adventure. What better than to mount a dinosaur hunt in the traditions of Arthur Conan Doyle and capture the ultimate proof for Queen and country?

The four-man team they gathered together included Mark Rothermel, Jonathan Walls and Joe Della-Porter. None was a scientist, however, which produced some problems: a few zoologists suggested this was not an authentic scientific quest. However, Professor Mackal was helpful. He was seeking the funding for another expedition himself. The plan was that the British would leave England in late 1985, departing the Congo in March 1986 just as Mackal and his team arrived. In this way the area would be occupied for many months with hopes of capturing concrete proof.

The Gibbons expedition tried to secure media backing and scientific funding but with limited success. The British supernatural magazine *Fortean Times* raised funds for the expedition and sold 'Operation Congo' T-shirts complete with sauropod logo.

Jenny spoke to Bill Gibbons just before he sailed from Liverpool. He was very confident, especially after persuading Dr Agnagna to go with them. The Brazzaville scientist was determined to overcome the ridicule he had received after his photographic fiasco and, as Gibbons pointed out, this meant that they had the best of both worlds – trained survivalists accompanied by the only scientist to have claimed a sighting of the dinosaur.

Unhappily, the expedition did not go well. The political situation caused problems and most of their money was spent paying a new government structure which had deposed the officials who first sanctioned the trip. Dr Agnagna also resigned and would not take them to the spot where he had filmed the monster three years before. As such,the expedition failed to capture a dinosaur, although it did find new eyewitnesses and some more footprints. The natives were so afraid of the beast that they would not take the Gibbons' team deeper into the jungle without the payment of large sums of money that they could not afford. Financial problems and the changed political situation also wrecked Mackal's new expedition before it began. The new government insisted that the creature be protected and that no attempt be made to do more than film it from a distance if it was discovered.

However the attraction of finding a real dinosaur is still considerable. In late 1992 Bill Gibbons concluded years of preparation and set off on Operation Congo 2 – a six-week reconnaissance trek into the Likouala swamps as a prelude to a future mission that will be mounted around a humanitarian supply run: in exchange for taking in medical supplies as well as badly needed food Gibbons was promised native cooperation and guides to the area where the *mokele mbembe* lives.

Upon his return to Britain in early 1993 he reported satisfaction with the plans and said that it all was set for a significant attempt to obtain proof of the dinosaur. The expedition is planned for spring 1994.

Monsters of the mind

There are, of course, people who remain very sceptical about the possibility of a living dinosaur. Loren Coleman, an intrepid seeker of mysterious animals followed up some of the Texas pterosaur sightings and concluded that a form of wish fulfillment was at work in some cases. There had been considerable publicity and speculation about the discovery of fossils of a new type of dinosaur. The subject was, therefore, of popular interest when the sightings occurred. This seems rather coincidental. It was proposed that perhaps large examples of known birds were being seen by people and, mindful of the dinosaur legends, they thought they had seen a *pteranodon*

Elsewhere the realities of the tourist industry also have to be taken into account. If a dinosaur is believed to live in an area, the locals will want to ensure that its legend survives for it may attract visitors, aid and scientific interest to their land.

Occasionally very peculiar stories emerge. For instance there has been a regular spate of sightings of pterodactyl-like creatures over the Yorkshire Moors in England. There were enough of these reports for them not to be completely dismissed and it was some time before a solution emerged. A group of model-makers had, in fact, built a remote-controlled swooping pteranodon for a low-budget science-fiction movie. Years after the shooting, this creature was still giving the odd private display around Ilkley Moor and terrifying one or two fell-walking observers.

Another intriguing idea suggested by some researchers is that dinosaur sightings might be similar to the video-replay type of ghost.

A reconstruction of what a living pteranodon would look like. Witnesses in Texas have reported encounters with flying creatures just like this

Parapsychologists wonder if signals trapped in certain rocks can, from time to time, replay scenes from the distant past. Instead of seeing a human figure the witness may observe a long-dead animal. This would certainly account for the fact that no bones, skeletons, carcasses or physical evidence has ever been found. Zoologists rightly point out that if real dinosaurs had survived such proof would surely be discovered occasionally. On the other hand, video projections could not possibly explain huge gaps smashed into tree foliage or native canoes overturned by surfacing monsters, such as the evidence dominating the Likoula swamp sightings.

In some cases at least it does seem possible that these creatures may exist in a distinctly physical sense.

Rebuilding dinosaurs

In recent years there have been ambitious ideas to rebuild dinosaurs as living animals in the laboratory. This far-fetched but controversial concept was used to great effect in 1993 in Steven Spielberg's special-effects blockbuster movie *Jurassic Park*. Here the 'model' dinosaurs are part of an island theme-park. (The Jurassic period occurred 193–136 million years ago in the Mesozoic Era.)

The concept has grown because of recent developments in genetic engineering. Scientists now understand the basic building blocks of life – that is the DNA molecules found within all creatures. It has, therefore, become possible to alter and adapt these to some extent and to encourage embryos to grow in very specific ways. The idea of cloning animals by taking

Dinosaur facts

■ Numerous species of dinosaurs were the dominant form of life on Earth between approximately 225 and 64 million years ago. They ruled land, sea and sky many times longer than human beings have so far held ascendancy on this planet.

■ Contrary to a widely popular myth, nothing even remotely like a human being ever co-existed with dinosaurs.

■ All remaining species died out over 60 million years ago in terms that geologists consider to be very sudden (although this may have taken at least several centuries to be completed in real time). The cause of this global extinction is still hotly debated and the dispute is between gradual changes (e.g. in the living environment) or, the more favoured, cataclysmic event (possibly an astronomical disaster that altered the Earth's climate almost overnight). Either way it could happen again at any time – to us!

■ Persistent reports suggest that very few dinosaurs might have survived and evolved through the eras in remote areas of the world, such as the deep oceans or the remote hinterland of the Australian bush.

■ The most widely sought creature is described by natives in terms that closely resemble an herbivorous dinosaur akin to, but smaller than, the *brontosaurus*. This has been reported as surviving in the intractable swamps of the African Congo during the twentieth century and several serious scientific expeditions have visited the area since 1975 attempting to capture proof of this animal. There have been a number of near misses and some cryptozoologists (scientists who seek undiscovered animals) think there is at least a fifty/fifty chance that this dinosaur will be proven to exist before the year 2000.

these cells and growing them under laboratory conditions to produce a series of identical animals is rightly causing ethical questions to be debated. If cells were ever obtained from a dinosaur perhaps it might be possible to genetically engineer an entire monster.

Some of the most interesting fossils to be found are of insects imbedded within resin. Tree sap oozed over them as they fed on a trunk and set hard trapping them in their entirety with skin, body tissues, blood and organs intact. When the resin solidified into amber over time this unique form of preservation was maintained.

Whole insects preserved from 30 million years ago have been studied and examples are known from the dinosaur era. It has been suggested that if a mosquito-like insect, that had sucked blood from a dinosaur, was killed immediately afterwards and preserved in amber, then cells from that dinosaur might remain in the insect's stomach and thus enabling the saurian DNA (that is self-replicating material) to be cloned. But as yet we do not have the technology to do this.

The end of the saurian world

One of the most fascinating riddles of modern scientific research is exactly why the dinosaurs died in such vast numbers and so very quickly. The disappearance did not occur, however, instantaneously as some fanciful ideas suggest. It is in geological terms that it was very quick being only a few centuries or even decades.

We know that something cataclysmic must have occurred about 64 million years ago from our study of the rocks in which fossils are found. One geological era ended and another began – something which took place on many occasions during the 5-billion-year history of the Earth. However, usually this change has been characterized by a gradual adaptation of the climate, making deserts bloom and seas dry up and thus changing the living conditions of the animals found there. As a result some forms of life fared well, others did not, and evolution ensured that only those best suited to the new environment survived into the next era.

However, the disappearance of thousands of species, including all of the saurians, was an unprecedented event. As the dinosaurs had dominated every sphere of life from the land to the air in the Mesozoic Era (225–64 million years ago), their extinction along with several other forms of life at the end of the Cretaceous Period requires an unusual explanation There have been many attempts to fathom out the reason for this including the idea that the Earth instantaneously flipped over in its orbit or reversed polarity. The more sedate theories –

Meteor crater, Arizona, the visible scar left on the earth's surface by a modestly sized chunk of rock from outer space which impacted before man lived on the planet. Did a much larger impact some 64 million years ago create such huge environmental disaster that all the earth's dinosaurs became extinct?

Authorities

■ *Loren Coleman* investigated the pterosaur episode in America with his colleague Jerome Clark. They adopted a careful, psychologically orientated approach. Their book *The Unidentified* (Warner, New York, 1975) is an interesting sideline. Coleman has continued to follow up monster stories in the USA and his book *Curious Encounters* published in 1985 is worth a look as a series of case investigations. Clark has also published reports on his strange animal research in *Unexplained!* (Visible Ink, Detroit, 1993)

■ *Bernard Heuvelmans* was a pioneer of cryptozoology. His book *On the Track of Unknown Animals* (Hart-Davis, New York, 1958) alerted many to the possible theories. He looked at delightful oddities such as the *tatzelwurm*, an unknown type of giant worm or lizard up to 30 feet (9 metres) long said to have been seen by climbers in the Alps in Austria and Switzerland.

■ *Professor Roy Mackal* is probably the best-known scientist today who is involved in expeditions for surviving dinosaurs. His book *Searching for Hidden Animals* (Doubleday, New York, 1980) describes his early explorations. His attempts to track down the *mokele mbembe* are discussed in detail in *A Living Dinosaur?* (E. J. Brill, New York, 1987)

■ *Karl Shuker* is a British zoologist who has made a particular study of animals in unexpected places and of preserved fossils. His most useful book is a collection of his papers, entitled *Extraordinary Animals Worldwide* (Robert Hale, London, 1991). His major opus is *The Collins Encyclopedia of New and Rediscovered Animals* (Harper Collins, 1994)

that mammals consumed dinosaur eggs or the living environment of the creatures became slowly intolerable – do struggle to explain the extent and relative suddenness of the dinosaurs' demise. However, the one catastrophic theory that most scientists take seriously is that the Earth was struck by a gigantic meteorite.

The Earth is constantly bombarded by pieces of rocks from space, much of it being left over debris from the time when the solar system formed billions of years ago. It lies in space like a vast junk yard and when the Earth's passage around the sun intercepts assorted debris patches every year some of it heads towards the surface causing a natural collision. But most of this debris is very small – grain-sized particles or rocks a few inches in diameter. These burn up by friction as they hit the thick barrier formed by the gasses in the upper atmosphere. We can see the result as a dim streak of meteoritic light in the sky on many nights. Occasionally the rock is bigger or has harder metallic ores inside that burn up less easily. These can hit the Earth causing craters. Such impacts result from fireball meteors or bolides and several are seen over various parts of the Earth every year – often being spectacular and long-lasting enough to provoke waves of UFO reports. Several spectacular meteorite impacts happen every century.

On extremely rare occasions meteors the size of houses or city blocks can bombard the atmosphere. Meteor Crater, Arizona, USA, is a huge hole like a gigantic quarry that was gouged out after one of the biggest known collisions several million years ago. It is a virtual certainty that some larger impacts must have taken place from time to time. A collision with a rock, one or two miles (1.5–3 km) in diameter is perfectly possible every few millenia.The consequences of this would be global rather than local. If such a rock struck an ocean huge tidal waves would swamp most coastal areas killing billions of creatures very quickly. Dust from such a meterorite would mask the sun out for years and water vapour thrown up would have trapped heat from the sun when the dust settled. Those animals not killed by the impact or its immediate consequences would probably succumb to the rapid climate change that would result either as the climate cooled with no sun and later as it warmed when the dust settled. Some species would come through relatively unscathed, such as insects or small mammals that could burrow underground or hide in caves for warmth. The flying, swimming lumbering dinosaurs, which were not noted for their environmental adaptability, would have found it hard. It is by no means inconceivable that such an impact would herald their complete if lingering death.

Traces of iridium – a rare substance on Earth but common in meterorites – have been found in a very thin layer at exactly the correct boundary between the geological Mesozoic and

Tertiary eras. This iridium layer is consistent with the fall to Earth of vaporized debris from a giant meteorite. This deadly snow would have covered much of the world's surface.

Such a disaster can strike at any time. It could be that creatures evolved from insects several million years from now will be speculating about the strange and sudden disappearance of a long-lost race of fearsome bipedal creatures who thought they ruled the world.

A scene from the Jurassic period when dinosaur species ruled the land, sea and air. The plesiosaur is thought by some cryptozoologists to survive in a few locations and be reported by terrified witnesses

SOURCES

As active field researchers much of the material in this volume is the result of our own private investigations into the realms of the unexplained. However, we also owe a debt to other researchers and writers and credit our main sources below.

Books and papers

Bardens, Dennis *Psychic Animals* (Robert Hale 1988)
Beer, Lionel *The Moving Statue of Ballinspittle* (Spacelink Books 1986)
Bord, Janet and Colin *Alien Animals* (Granada 1980)
Blackmore, Dr Susan *Dying To Live* (HarperCollins 1993)
Bright, Michael *There are Giants in the Sea* (Robson Books 1989)
Brookesmith, Peter (editor) *The Unexplained* (Orbis part-work 1980–3)
Campbell, Steuart *Loch Ness Monster – The Evidence* (Aquarian 1986)
Cavendish, Richard (editor) *Man, Myth & Magic* (Purnell part-work 1970–2)
Coleman, Loren and Clark, Jerome *The Unidentified* (Warner 1975)
Coleman, Loren *Curious Encounters* (Warner 1985)
Cooper, Joe *The Case of the Cottingley Fairies* (Hale 1990)
Delgado, Pat and Andrews, Colin *Circular Evidence* (Bloomsbury 1989)
Delgado, Pat *Crop Circles: Conclusive Evidence?* (Bloomsbury 1991)
Dinsdale, Tim *The Leviathans* (Routledge & Kegan Paul 1966/Futura 1976)
Dinsdale, Tim *Project Water Horse* (Routledge & Kegan Paul 1975)
Evans, Hilary *Visions, Apparitions, Alien Visitors* (Aquarian 1986)
Fairley, John and Welfare, Simon *Arthur C. Clarke's Chronicles of the Strange and Mysterious* (Collins 1987)
Frank, Dr Louis *The Big Splash* (Avon 1991)
Grey, Dr Margo *Return from Death* (Arkana 1986)
Grumley, Michael *There are Giants in the Earth* (Sidgwick & Jackson 1975)
Heuvelmans, Bernard *On the Track of Unknown Animals* (Hart-Davis 1958)
Hough, Peter *Witchcraft – A Strange Conflict* (Lutterworth 1991)
Keel, John *Our Haunted Planet* (Futura 1975)
Kubler-Ross, Elisabeth *On Death And Dying* (MacMillan 1969)
Mackal, Professor Roy *Searching For Hidden Animals* (Doubleday 1980)
Mackal, Professor Roy *A Living Dinosaur?* (E. J. Brill 1987)
Meaden, Dr Terence *The Circle Effect and Its Mysteries* (Artetech 1989)
Meaden, Dr Terence *Goddess of the Stones* (Souvenir 1991)
Meaden, Dr Terence and Elsom, Dr Derek (editors) *Circles from the Skies* (Souvenir 1991)
Meurger, Michael with Gagnon, Claude *Lake Monster Traditions* (*Fortean Tomes* 1988)
Michell, John and Rickard, Robert J. M. *Phenomena – A Book of Wonders* (Thames & Hudson 1983)
Moody, Dr Raymond *Life after Life* (Bantam 1975)
Moody, Dr Raymond *Reflections on Life after Life* (Bantam 1977)
Moody, Dr Raymond *The Light Beyond* (Bantam 1988)
Morse, Dr Melvin *Close to the Light* (Souvenir 1991)
Noyes, Ralph (editor) *The Crop Circle Enigma* (Gateway 1990)
Persinger, Dr Michael *Space-Time Transients and Anomalous Phenomena* (Prentice–Hall 1977)
Randles, Jenny and Hough, Peter *Scary Stories* (Futura 1991)
Randles, Jenny *Sixth Sense* (Hale 1987)
Randles, Jenny and Hough, Peter *Death By Supernatural Causes?* (Grafton 1988)
Randles, Jenny and Hough, Peter *Spontaneous Human Combustion* (Hale 1992)
Randles, Jenny and Fuller, Paul *Crop Circles – A Mystery Solved* (Hale 1993)

Ring, Dr Kenneth *Life at Death* (McCann & Geohegan 1980)
Ring, Dr Kenneth *Heading Towards Omega* (Morrow 1984)
Ring, Dr Kenneth *The Omega Project* (Morrow 1992)
Sabom, Dr Michael *Recollections of Death* (Harper & Row 1982)
Searle, Frank *Seven Years in Search of the Monster* (Coronet 1976)
Shuker, Karl *Extraordinary Animals Worldwide* (Hale 1991)
Shuker, Karl *The Collins Encyclopedia of New and Rediscovered Animals* (HarperCollins 1994)
Sieveking, Paul *Fortean Times – Diary of a Mad Planet* (*Fortean Tomes* 1991)
Schul, Bill *The Psychic Frontiers of Medicine* (Ballantine 1977)
Spencer, John and Evans, Hilary (editors) *Phenomenon* (Futura 1988)
Taylor, John *Science and the Supernatural* (Temple Smith 1980)
Tributsch, Dr Helmut *When The Snakes Awake* (MIT Press 1982)
Tomas, Andrew *Beyond The Time Barrier* (Sphere 1974)
Vallée, Jacques *Passport To Magonia* (Spearman 1970)
Watson, Dr Lyall *The Nature of Things* (Hodder and Stoughton 1990)
Wentz, W. Y. Evans *The Fairy Faith in Celtic Countries* (Smythe 1981)
Witchell, Nicholas *The Loch Ness Story* (Corgi 1989)

Radio and television documentaries

Super Powers Equinox series (Channel 4, 18 November 1990)
The Great Sea Monster Mystery presented by Fergus Keeling (BBC Radio 4, December 1986)

Magazines and journals

'Psychic Healing in Danbury, Essex' *Cosmology Newslink* Spring 1991
'Searching for the Historical Bigfoot', Michael T. Shoemaker, *Strange Magazine* No. 5
'Seeing Red', Bob Rickard, *Fortean Times* Issue 65, Oct/Nov 1992
'Stella Lansing's Clocklike UFO Patterns', Dr Berthold Eric Schwarz *Flying Saucer Review* Vol. 21 No. 1 June 1975

PICTURE ACKNOWLEDGMENTS

The publishers are grateful to the following for permission to include the pictures reproduced on the pages indicated:

Michael Bromley: p.109

The Fortean Picture Library: pp. ii, xiv, 2, 4, 5, 7, 8, 22, 25, 27 (above), 33, 40, 42, 43, 44, 48, 49, 53, 54, 57, 60, 61, 65, 66, 69, 98, 101, 114, 122, 126, 129, 130, 131, 133, 134, 136, 139, 141, 144, 147, 148, 157, 160 (below), 167, 171, 183, 184, 187, 190, 193, 194, 197, 199, 201, 203, 204, 206, 207, 209, 211, 212, 214, 215, 216, 217, 218, 219, 225, 229. Plate section, pp.1, 4, 6, 8

John Frost, Historical Newspaper Service: pp.176-7

Peter Hough: p.97, 164, 165, 168

The Mansell Collection: 80, 82, 113

Mary Evans Picture Library: 38, 63, 85, 93, 185. Plate section, pp. 2, 5, 7

Mary Evans Picture Library / G. L. Playfair: p.20

Mary Evans Picture Library / Harry Price: p.162

Tony McMunn: plate section, p.3

Siegfried Pracher / PSI Research Archive: p.112

Jenny Randles: pp.17, 27 (below), 77

Royal Geographical Society: p.186 (photo: Eric Shipton)

Roy Sandbach: p.13

Syndication International: pp. 88, 107

Topham: pp.15, 55, 106, 115, 120, 163, 188, 208, 220, 222, 227

INDEX